Geologic Analysis of Naturally Fractured Reservoirs

Geologic Analysis of Naturally Fractured Reservoirs

Editor

Govind Kumar

scitus
academics

Geologic Analysis of Naturally Fractured Reservoirs
Edited by **Govind Kumar**

Printed in 2017

ISBN: 978-1-68117-389-4

Library of Congress Control Number: 2015941578

SCITUS Academics LLC,
616, Corporate Way, Suite 2, 4766,
Valley Cottage, NY 10989

www.scitusacademics.com

Notice

Contents

Preface

Geologists, engineers, and petrophysicists concerned with hydrocarbon production from naturally fractured reservoirs. In this book, we discusses a valuable tool for obtaining pertinent rock data to evaluate reserves and optimize well location and performance. Indeed, from exploration through blow-down, the effective management of these reservoirs requires the application of multi-disciplinary approaches more than virtually any other type of petroleum reservoir. Geologists, engineers, and petrophysicists concerned with hydrocarbon production from naturally fractured reservoirs will find a valuable tool for obtaining pertinent rock data to evaluate reserves and optimize well location and performance.

Editor

Chapter 1

Sustainable Heat Farming: Modeling Extraction and Recovery in Discretely Fractured Geothermal Reservoirs

Don B. Fox[a, b, c], Daniel Sutter[d], Koenraad F. Beckers[a, b, c], Maciej Z. Lukawski[a, b, c], Donald L. Koch[a, b, c], Brian J. Anderson[e], and Jefferson W. Tester[a, b, c]

[a]School of Chemical and Biomolecular Engineering, Cornell University, Ithaca, NY 14853, USA

[b]Cornell Energy Institute, Cornell University, Ithaca, NY 14853, USA

[c]Atkinson Center for a Sustainable Future, Cornell University, Ithaca, NY 14853, USA

[d]Institute of Process Engineering, ETH Zurich, Zurich, Switzerland

[e]Department of Chemical Engineering, West Virginia University, Morgantown, WV 26506, USA

ABSTRACT

Although many natural hydrothermal geothermal systems have been shown to be productive over long periods of time, limited field testing of Enhanced or Engineered Geothermal Systems (EGS) has prevented adequate assessment of their sustainability. To estimate how renewable EGS reservoirs might be, an analytical approach employing Green's function was used to model transient thermal conduction in an idealized reservoir containing a single rectangular fracture to evaluate heat transfer effects during alternating periods of extraction and recovery. During recovery, the temperature along the fracture surface approaches the temperature of the bulk rock with the deviation from the surrounding bulk temperature decaying as $1/\sqrt{t}$ where t is the recovery time. Numerical simulations of a multiple parallel fracture reservoir using the TOUGH2 code agreed with the derived analytical solutions over a range of flow rates and interfracture spacings with only small deviation due to multidimensional effects. Multidimensional effects are more pronounced near the inlet and outlet of the fracture and are reduced at higher flow rates. Thermal interactions between sufficiently spaced fractures are negligible for production periods of 10–30 years, suggesting that the single fracture analytical model can be applied to multifracture reservoirs provided that the mass flow used is on a per fracture basis. Simulation results show that multifracture EGS reservoirs have a greater capacity to sustain high outlet temperatures, suggesting that conductively dominated EGS systems can be regarded as renewable over time scales of societal utilization systems (three to five times the heat extraction time).

INTRODUCTION

The practical utilization of geothermal energy requires extracting or mining the stored thermal energy contained in the Earth's crust, often several kilometers deep. Today, geothermal energy reservoirs exploit naturally occurring hydrothermal aquifers with high permeability and porosity and sufficient quantities of water or steam. Unfortunately, nature does not always provide these ingredients in locations with large temperature gradients. Engineered or Enhanced Geothermal Systems (EGS) offer technologies that can be implemented in areas with low permeable and/or porous rock formations that do not contain enough water and/or steam or lack sufficient interwell connectivity. In most EGS reservoirs, the rock formations must be stimulated, typically using hydraulic pumping resulting in fracturing or hydroshearing to open pre-exiting sealed fractures. In this manner, the reservoir is engineered, creating open fractures and connectivity where liquid water can then flow under increased pressure through the fracture system to extract heat by contacting regions of hot rock.

One important feature of any operating geothermal reservoir system is its anticipated production sustainability over the long term. In simple terms, the heat mining process removes the stored thermal energy faster than it can be supplied by steady state conduction. Although geothermal reservoirs will be depleted during production, with proper management, hydrothermal reservoirs have been shown to be productive over long periods of time (Axelsson et al., 2005 and Bromley et al., 2006). The situation is different for EGS as there is no record of long term field testing and as a result, the renewability of EGS is often questioned. Axelsson et al. (2001) define the renewability of a geothermal resource as the ability to maintain the installed capacity indefinitely. Strictly speaking, to maintain capacity indefinitely, the reservoir would have to operate in steady state where the rate of heat extraction from the reservoir is equal to the rate of recharge of thermal energy coming back into the reservoir from the surrounding bulk rock. Unfortunately, due to practical requirements for production, plant needs, and operation costs, the renewable capacity is often too small for commercial development (Sanyal, 2005). Therefore, considerations about the sustainability of geothermal systems must also include the thermal recovery that occurs during a pause in heat extraction. Megel and Rybach (2000) explore the sustainability of a low temperature (62 °C) hydrothermal doublet used for district heating. By allowing for thermal recovery after a period of thermal extraction, they concluded that because the hydrothermal doublet renews itself on a time scale comparable to extraction, it can be operated in a sustainable manner. Sanyal (2005) defines the sustainable capacity of a geothermal reservoir as the capacity that can be economically maintained over the amortized life time of a power plant. In this study, we attempt to quantify the recovery of the reservoir on an absolute basis in terms of its thermal productivity irrespective of any power plant or surface installation life time considerations. Note that unfortunately, recovery is sometimes used as a synonym to thermal extraction (e.g., Elsworth, 1989) in the sense that thermal energy is recovered from the subsurface for use in the surface installation. To avoid confusion, in this study extraction is used to denote the process of removing heat from the subsurface while recovery describes the process of heat from the surrounding hot rock transferring back into the reservoir after extraction has stopped.

Although exploiting an EGS reservoir at an economically useful rate eventually results in significant local cooling of the reservoir, there are valid reasons to consider such geothermal resource operations as sustainable. Complete recovery of the thermal energy is eventually guaranteed following the discontinuation of all production as heat is conducted and convected to the surrounding hot rock that contains the active reservoir. Even with recovery times of hundreds of years, the resource base is large enough to allow for long term sustainable energy production on a national or global scale (Tester et al., 2006). Additionally, geothermal resources can be considered renewable on time scales of technological/societal systems, 10–100 years, whereas fossil fuel reserves "renew" on geologic time scales of order hundreds of million years (Rybach et al., 2000). Furthermore, displacement of fossil fuel consumption with geothermal or other renewables would reduce environmental pollution and help to preserve hydrocarbons for future generations as raw materials and chemical feedstocks as well as for cleaner and more efficient power applications (Sanyal, 2005).

To guarantee the long term sustainability and renewability of a specific EGS reservoir or field development would require implementing a heat farming strategy. Under this approach, several

EGS reservoirs would be used in a rotating manner. When the first reservoir is "depleted" to a point where it is no longer able to satisfy the thermal needs of the surface installation, production would be shifted to a new reservoir. The process would continue until the first reservoir has recovered sufficiently to be restarted or restimulated.

The main objective of this work is to evaluate how effectively model geothermal reservoirs recover following extraction. An analytical and numerical model were used to characterize the transient thermal hydraulic behavior of a fractured EGS reservoir. The analytical model considers an idealized system with single and multiple parallel isolated fractures embedded in a large body of impermeable rock with uniform properties. In this situation, heat transfer is dominated by transient conduction perpendicular to the fracture surface coupled to convection of heat by the fluid flowing along the fracture. The TOUGH2 numerical code developed at the Lawrence Berkeley National Laboratory (Pruess et al., 1999) was used for modeling multiple parallel fractures with two-dimensional transient heat conduction within the rock to test the assumptions of the analytical model.

Although the parallel multi-fracture system is highly idealized and is impossible to replicate in practice, such a simple model was chosen to capture the basic behavior of a conduction dominated EGS reservoir undergoing alternating phases of extraction and recovery. This behavior is relevant to evaluating the sustainability of EGS and has been virtually ignored by other researchers. Although our model only represents thermal modeling of a very idealized, conduction dominated EGS reservoir, we feel it illustrates the limiting behavior one might expect in practice – therefore providing a conservative evaluation of sustainability. Real EGS reservoirs will, of course, contain nonlinear coupled thermal, hydraulic, mechanical, and chemical (THMC) processes. However, introducing these features into the analytical model would not result in tractable solutions relevant to the main objective of the paper.

Much research has been done in addressing nonlinear and coupled phenomena in EGS reservoirs. Kohl et al. (1995) modeled coupled THM processes in a single fracture using linear thermo-poroelastic effects and a logarithmic closer law and learned that these effects lead to significant long term reduction in impedance of flow. Willis-Richards and Wallroth (1995) summarized the current EGS modeling at the time and emphasized the importance of accurately capturing the evolution of aperture near the injection and production well, large scale permeability enhancement through means of fracture extension, and heat conduction in the reservoir. Taron and Elsworth (2009) studied THMC coupling effects in a dual porosity media where porosity and permeability change as a result of change of stress and chemical precipitation and dissolution. Their numerical simulations indicate that mechanical effects dominate in the short term (days), while thermal effects are important in the intermediate term (months) and chemical effects are only important in the long term (years). Simulations also reveal that coupling of these effects are needed to explain observed phenomena that parallel reactive transport and geomechanic simulations cannot capture. Ghassemi and Zhou (2011) coupled fracture flow and heat transport to thermo-poroelastic deformation in a discretely fractured reservoir and found the resulting temperature, pressure, stress, and aperture at every time step. Their numerical model revealed that poroelastic effects are dominant at early stages while thermoelastic effects take over at later time and coincide with less leakoff to the far field and significant increase in fracture aperture.

ANALYTICAL MODEL

A representation of a parallel discretely fractured EGS reservoir, adapted from earlier works by Gringarten et al. (1975) and Wunder and Murphy (1978), is pictured in Fig. 1. For a single fracture model, the fracture spacing x_s goes to infinity. Although realistic reservoirs consist of a network of fractures, a single fracture model adequately captures the essential features of thermal extraction through heat conduction from the rock surrounding the fracture. A single rectangular, vertical fracture of constant width $2b$ separates two blocks of homogeneous, isotropic, impermeable rock. The orientation of the fracture does not matter because buoyancy driven flow is not considered; however, hydraulic stimulation of a reservoir will usually yield vertically oriented fractures because fractures propagate in the direction normal to the plane of least principal stress. A Cartesian coordinate system has been placed such that x, $z = 0$ coincides with the inlet.

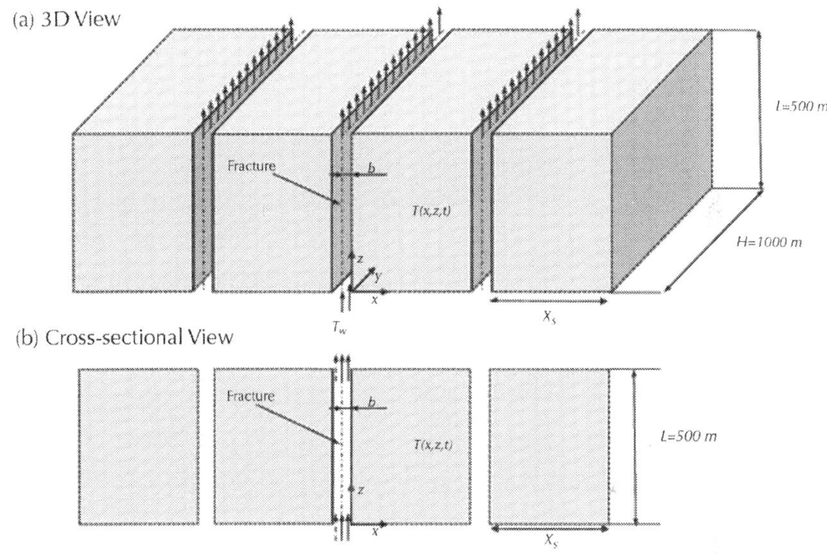

Figure 1: Conceptualized multifracture EGS reservoir. (a) Three-dimensional view of the multifracture EGS reservoir. x_s represents the spacing in between fractures. (b) Reduced two-dimensional view.

In the isolated fracture model, the rock is assumed to extend in x to $\pm\infty$. Both the rock and the fracture extend to infinity in the y direction. Initially ($t \leq 0$), this semi-infinite system is at a uniform temperature T_r. During the heat extraction phase, water is injected at x, $z = 0$ for all y at a constant temperature Tw and a constant mass flow rate. The water flows parallel through the fracture to the outlet at $x = 0$ and $z = L$. Furthermore, the following assumptions are made:

- The temperature variation in x in the water is negligible, as the aperture of the fracture is very small compared to the diffusion length in x $(b << \sqrt{\alpha t})$. For all z, the water temperature is equal to the rock temperature at $x = 0$. In other words, all heat transfer resistance at the rock/water interface is neglected, which a reasonable assumption for any practical case is given the low thermal conductivity of the rock (McFarland and Murphy, 1976, Wunder and Murphy, 1978 and Ogino et al., 1999).

- Conduction in the vertical direction, z, in both the fracture and the rock formation is neglected. Because the aperture $2b$ of the fracture is small compared to the fracture extent in the y direction H, such that $2b/H \ll 1$, we can extend the y axis to $\pm\infty$ and ignore the heat transfer in the y direction, effectively reducing the heat transfer problem to two spatial dimensions. Heat transfer occurs only by conduction in the rock in the x direction and forced convection along the z direction in the fracture. Radiative heat transfer is also neglected.
- No heat flux occurs at the rock/water interface during recovery when no fluid is circulating. This assumption is justified by the small volume, and hence, small overall heat capacity of the fluid filling the fracture compared to the total heat capacity of the rock that has been cooled during the preceding extraction phase.
- The static fluid pressure in the fracture is set to exceed the vapor pressure of the water by a large enough margin to sustain single phase flow in the fracture. The pressure drop due to viscous forces is small and results in negligible temperature rise.
- The thermophysical properties of both water and rock are constant and independent of temperature and position.

First Extraction Phase

A one-dimensional differential energy balance within the rock yields

$$\frac{\partial T}{\partial t} - \alpha \frac{\partial^2 T}{\partial x^2} = 0 \tag{1}$$

Where T is the temperature and α is the rock thermal diffusivity, i.e., the ratio of its thermal conductivity k_r to the product of density ρ_r and specific heat capacity $c_{p,r}$. By applying a steady state energy balance to the water in the fracture over a differential length in z, the boundary condition for the rock/water interface is

$$\left(\rho_w c_{p,w} U b \frac{\partial T}{\partial z} = k_r \frac{\partial T}{\partial x} \right) \bigg|_{x=0} \tag{2}$$

With U representing the flow velocity of the water, b the half width of the fracture, and ρw and $c_{p,w}$ the density and specific heat capacity of the water, respectively. With the initial and bulk temperature of the rock specified as T_r and the constant water temperature at the injection position taken as T_w, the additional boundary conditions are

$$T(0, 0, t) = T_w, \quad t > 0, \tag{3}$$

$$\lim_{x \to \infty} T(x, z, t) = T_r, \tag{4}$$

And the initial condition is

$$T(x, z, t \leq 0) = T_r. \tag{5}$$

The analytical solution is given in terms of the dimensionless, normalized temperature Θ and is based on the similarity solution of a classical transient heat transfer problem as presented by Carslaw and Jaeger (1959) and Arpaci (1966):

$$\Theta \equiv \frac{T(x, z, t) - T_w}{T_r - T_w} = \text{erf}\left[\frac{x + \beta z}{2\sqrt{\alpha t}} \right], \tag{6}$$

Where β is defined as

$$\beta \equiv \frac{2k_r}{(\dot{m}/A)c_{p,w}L}. \tag{7}$$

Here, m/A is the mass loading of the fracture where *A* is defined as the one face area of the fracture and equals *HL* and *L* is the length of the fracture parallel to the flow. In terms of a temperature difference with respect to the bulk rock, the temperature profile in the reservoir is equal to

$$u(x, z, t) = u_0 \mathrm{erfc}\left[\frac{x+\beta z}{2\sqrt{\alpha t}}\right], \tag{8}$$

Where $u \equiv T(x, z, t) - T_r$ and u_0=Tw−Tr. The relationship between u and Θ is thus $\Theta = 1 - u/u_0$. Eq. (6) can be evaluated for the temperature at the outlet, $x = 0$ and $z = L$, (for a fracture of length L) and rewritten in a more familiar form as

$$\Theta(0, L, t) = \mathrm{erf}\left[\left(\frac{\dot{m}}{A}\right)^{-1}\frac{1}{c_{p,w}}\sqrt{\frac{k_r \rho_r c_{p,r}}{t}}\right]. \tag{9}$$

Validity of 1-D Conduction

The solution for the temperature field in the previous section neglects heat conduction parallel to the fracture. Conventional scaling analysis can be used to compare the order of magnitude of heat conduction in the two directions and to assess the assumption of negligible heat conduction in z. The terms in the governing equation for two-dimensional differential energy balance scale as

$$\frac{\partial^2 T}{\partial x^2} \sim \frac{\Delta T}{\delta_x^2} \quad \text{and} \quad \frac{\partial^2 T}{\partial z^2} \sim \frac{\Delta T}{\delta_z^2}, \tag{10}$$

Where ΔT, δ_x, and δ_z are the appropriate characteristic quantities for temperature and length in *x* and *z*, respectively. Hence, we can define a dimensionless quantity D as

$$\mathcal{D} \equiv \frac{\text{heat conduction in } z}{\text{heat conduction in } x} = \left(\frac{\delta_x}{\delta_z}\right)^2 \tag{11}$$

In order to determine the relative importance of heat conduction in each direction, the characteristic length scales in both the *x* and *z* direction must be determined. δ_x and δ_z can be considered to be the penetration lengths of the thermal cooling front along the *x* and *z* axis, respectively. Defining the position of the thermal cooling front to be the value of *x* or *z* at which Θ equals an arbitrary value, the thermal penetration lengths are

$$\delta_x \sim \sqrt{\alpha t} \quad \text{and} \quad \delta_z \sim \frac{1}{\beta}\sqrt{\alpha t}. \tag{12}$$

Eq. (12) takes the typical form of penetration lengths of conduction dominated problems. From Eq. (12), our dimensionless quantity scales as follows

$$\mathcal{D} \sim \beta^2 \propto \left(\frac{A}{\dot{m}}\right)^2. \tag{13}$$

For m=70 kg/s, single face fracture area *HL* of 500,000 m^2 and typical thermophysical properties of water and rock, $\beta = 0.02$ and $\beta^2 = 4 \times 10^{-4}$. Thus, for typical flow rates, heat conduction in *x* is three orders of magnitude greater than in *z*, validating the approximation of neglecting heat conduction in *z*. The factor β^2 acts as a perturbation parameter in the full form of Eq. (1) and the solution presented in Eq. (8) is the *O* (1) solution. Because there is no convective transport of heat in the *z* direction below the inlet and above the outlet, a boundary layer (in *z*) of thickness $\varepsilon \sim \sqrt{\alpha t}$ is present in these regions. As a result of the boundary layers, Eq. (8) solution is most accurate for *z* values between □ and *L* – □.

First Recovery Phase

An approach using Green's functions was used to describe the temperature in the single fracture model analytically. Green's functions provide a systematic way to solve linear partial differential equations with inhomogeneous terms. We consider a linear heat sink along the fracture, i.e., for all z at $x = 0$. The line sink represents the heat extraction by the flowing fluid in the fracture during the extraction phase and varies in both z and t. The energy balance including the heat sink becomes

$$\frac{\partial T}{\partial t} - \alpha \frac{\partial^2 T}{\partial x^2} = \frac{q(z,t)\delta(x)}{\rho_r c_{p,r}} \tag{14}$$

Where $q\ (z, t)$ is the net heat flux from the fracture to the rock (which is typically negative) and $\delta(x)$ is the Dirac delta function. As mentioned before, the temperature at $x = 0$ is valid for both rock and water at the interface. We are now solving the problem in unbounded space with the initial condition for Eq. (14)again represented by Eq. (5) and two boundary conditions that the temperature goes to T_r at $x = \pm\infty$.

During the extraction phase, heat is conducted from the hot regions of rock into the moving fluid in the fracture at $x = 0$. The flux can be determined based on the temperature profile for the extraction in Eq. (6). No heat flux occurs at the rock/water interface during the recovery phase and thus $q\ (z, t)$ is identically zero. With t_1 being the time at the end of the extraction phase, we get

$$q(z,t) = \begin{cases} -2k_r \left.\dfrac{\partial T}{\partial x}\right|_{x=0}, & 0 < t \le t_1 \\ 0, & t > t_1. \end{cases} \tag{15}$$

The temperature solution during the first extraction, Eq. (8), is used to evaluate the flux in Eq. (15). The factor of two takes into account the two faces of the fracture. Eq. (14) was solved using the variable transformation u and the following Green's function (Duffy, 2001)

$$G(x - x', t - t') = \frac{H(t - t')}{\sqrt{4\pi\alpha(t - t')}} \exp\left[-\frac{(x - x')^2}{4\alpha(t - t')}\right], \tag{16}$$

Where $H(t - t)$ is the Heaviside step function, which is equal to 1 for positive arguments and equal to 0 for negative arguments. For details on the derivation, see Appendix A. The solution in integral form is

$$u(x,z,t) = \frac{u_0}{\pi}\int_0^{t_1} \frac{1}{\sqrt{t'}\sqrt{t-t'}} \exp\left[-\frac{1}{4\alpha}\left\{\frac{x^2}{(t-t')} + \frac{\beta^2 z^2}{t'}\right\}\right] dt'. \quad (17)$$

The integral in Eq. (17) has a closed form solution at the injection point x, $z = 0$ during the recovery ($t > t_1$):

$$\Theta(0,0,t) = \frac{2}{\pi}\tan^{-1}\left(\sqrt{\frac{t}{t_1} - 1}\right). \quad (18)$$

The solution in Eq. (18) was derived earlier by Wunder and Murphy (1978). Using a Taylor series expansion about 0 and termwise integration, the integral can be solved analytically for $x = 0$, i.e., for all positions along the fracture. The resulting infinite series is

$$\frac{u(0,z,t)}{u_0} = \frac{1}{\pi}\sum_{n=0}^{\infty}\binom{n-(1/2)}{n}\left(\frac{t_1}{t}\right)^{n+(1/2)}$$
$$\times\left[\xi_r^{2n+1}\frac{(-1)^{n+1}2^{n+1}\sqrt{\pi}}{(2n+1)!!}\operatorname{erfc}(\xi_r) + \right.$$
$$\left.\exp(-\xi_r^2)\times\sum_{k=0}^{n}\frac{(-1)^k 2^{k+1}\xi_r^{2k}}{(2n+1)(2n-1)\cdots(2n-2k+1)}\right] \quad (19)$$

Where $\xi_r = \beta z/(2\sqrt{\alpha t_1})$ and $\binom{n-(1/2)}{n}$ is a binomial coefficient. Eq. (19) only applies to the recovery phase and the rock/fracture interface, i.e., for $t > t1$ and $x = 0$. When the recovery time becomes large compared to the extraction time, t/t1 1, the heat sink due to the extraction fluid resembles an impulsive point line sink at $t = 0$ and $x = 0$. The first term of the infinite series in Eq. (19) is

$$u^{(0)}(0,z,t) = \frac{2}{\pi}u_0\left(\frac{t_1}{t}\right)^{1/2}\left[\exp(-\xi_r^2) - \sqrt{\pi}\xi_r \operatorname{erfc}(\xi_r)\right]. \quad (20)$$

Eq. (20) is equivalent to an impulsive point source at $t = 0$ and $x = 0$ of magnitude

$$Q(z) = \int_0^{t_1} q(z, t)\mathrm{d}t = 4k_r u_0 \left(\frac{t_1}{\pi\alpha}\right)^{1/2} \left[\exp(-\xi_r^2) - \sqrt{\pi}\xi_r \mathrm{erfc}(\xi_r)\right], \tag{21}$$

As Eq. (20) can be rewritten as

$$u^{(0)}(0, z, t) = \frac{Q(z)}{2\sqrt{\pi k_r \rho_r c_{p,r}}} \frac{1}{\sqrt{t}}. \tag{22}$$

If the recovery time is much larger than the period of the first extraction, then the temperature field depends only on the total heat extracted and not the temporal variations in the rate of extraction. Eq. (20) indicates that the temperature at the surface of the fracture will recover arbitrarily close to the initial rock temperature after a sufficient period of time. However, it also indicates that the deviation from the initial rock temperature retains a signature of the total heat extracted that is never forgotten. If multiple extraction phases were conducted, the longtime recovery temperature would depend on the total heat extracted in all the extraction phases.

Second Extraction Phase

For a geothermal reservoir undergoing three operational phases: extraction, recovery, and extraction, the heat flux at the rock/water interface ($x = 0$) is

$$q(z, t) = \begin{cases} -2k_r \left.\dfrac{\partial u}{\partial x}\right|_{x=0}, & 0 < t \le t_1 \\ 0, & t_1 < t \le t_2 \\ -2k_r \left.\dfrac{\partial u}{\partial x}\right|_{x=0}, & t > t_2, \end{cases} \tag{23}$$

Where t_2 is the end of the first recovery phase. Using the Green's function and applying the sifting property of the Dirac delta function and the Heaviside step function, the temperature field is

$$u_3(x, z, t) = \int_0^t \frac{\exp[-x^2/(4\alpha(t-t'))]}{\sqrt{4\pi\alpha(t-t')}} \frac{q(z, t')}{\rho_r c_{p,r}} dt'. \tag{24}$$

The subscript 3 in u_3 represents the temperature field during second extraction. Combining Eq. (23) with Eq. (24) yields

$$u_3(x, z, t) = \int_0^{t_1} \frac{\exp[-x^2/(4\alpha(t-t'))]}{\sqrt{4\pi\alpha(t-t')}} \frac{q(z, t')}{\rho_r c_{p,r}} dt' + \int_{t_2}^{t} \frac{\exp[-x^2/(4\alpha(t-t'))]}{\sqrt{4\pi\alpha(t-t')}} \frac{q(z, t')}{\rho_r c_{p,r}} dt'. \tag{25}$$

Since the first integral represents the integral solution for recovery with a duration of first extraction t_1, Eq. (25) can be rewritten as

$$u_3(x, z, t) = u_2(x, z, t; t_1) + \int_{t_2}^{t} \frac{\exp[-x^2/(4\alpha(t-t'))]}{\sqrt{4\pi\alpha(t-t')}} \frac{q(z, t')}{\rho_r c_{p,r}} dt'. \tag{26}$$

At $x = 0$,

$$\left(k_r \frac{\partial u_3}{\partial x} = \rho_w c_{p,w} U b \frac{\partial u_3}{\partial z} \right) \Bigg|_{x=0} \tag{27}$$

Which results in

$$u_3(x, z, t) = u_2(x, z, t; t_1) - \frac{1}{\beta}\sqrt{\frac{\alpha}{\pi}} \times \int_{t_2}^{t} \frac{\exp[-x^2/(4\alpha(t-t'))]}{\sqrt{(t-t')}} \frac{\partial u_3}{\partial z}(0, z, t')dt'. \tag{28}$$

By only considering points along the fracture ($x = 0$), Eq. (28) simplifies to

$$u_3(0, z, t) = u_2(0, z, t; t_1) - \frac{1}{\beta}\sqrt{\frac{\alpha}{\pi}} \int_{t_2}^{t} \frac{1}{\sqrt{(t-t')}} \frac{\partial u_3}{\partial z}(0, z, t')dt'. \tag{29}$$

Eq. (29) can be solved by applying Laplace transforms. The solution for $u_3\,(0, z, t)$ becomes

$$u_3(0,z,t) = u_1(0,z,t''') + u_2(0,z,t;t_1) + \frac{\beta}{\sqrt{\pi\alpha}} \int_0^{t'''} \int_0^{z} \left(\xi^2 - \frac{1}{2}\right) \frac{\exp(-\xi^2)}{\tau^{3/2}} u_2(0,z',t-\tau)dz'd\tau \tag{30}$$

With

$$\xi = \frac{\beta(z-z')}{2\sqrt{\alpha\tau}} \tag{31}$$

And

$$t''' = t - t_2. \tag{32}$$

u_1 represents the first extraction phase given by Eq. (8). The result indicates that the temperature during the second extraction is influenced by a delayed "fresh" extraction (i.e., a new extraction starting at $t = 0$ that proceeds as if no previous heat transfer had occurred), a continuing recovery, plus a convolution term involving the recovery temperature field. The convolution term represents the effect on the temperature field caused by the change in heat extraction rates during second extraction because the temperature field was not fully recovered before the extraction began. Thus, if the reservoir fully recovered, both the convolution term and u_2 would go to zero and Eq. (29) would reduce to Eq. (8), the solution for the first extraction.

Subsequent Extraction and Recovery Phases

One can follow a procedure analogous to that used in deriving the first recovery and second extraction phase to determine the temperature field for any extraction and recovery phase. For subsequent extractions, by using the Green's function, the appropriate piecewise heat flux $q(z, t)$ akin to Eqs. (15) and (23), and by recognizing integral forms of the temperature field of previous operational phases, one arrives at

$$u_n(0,z,t) = u_{n-1}(0,z,t) - \frac{1}{\beta}\sqrt{\frac{\alpha}{\pi}} \int_{t_{n-1}}^{t} \frac{1}{\sqrt{(t-t')}} \frac{\partial u_n}{\partial z}(0,z,t')dt'. \tag{33}$$

Odd values of n represent extraction phases while even values of n represent recovery phases. The parameter tn represents the time at the end of the nth phase. As was done for the second extraction, Laplace transforms are used to solve Eq. (33). The result is

$$u_n(0,z,t) = u_1(0,z,t^{(n)}) + u_{n-1}(0,z,t) + \frac{\beta}{\sqrt{\pi\alpha}} \int_0^{t^{(n)}} \times \int_0^{z} \left(\xi^2 - \frac{1}{2}\right) \frac{\exp(-\xi^2)}{\tau^{3/2}} u_{n-1}(0,z',t-\tau)dz'd\tau. \tag{34}$$

where $t^{(n)} = t - t_{n-1}$ and $t^{(n)}$ represents a reset time such that time zero occurs at the beginning of the nth operational phase. Similar to extraction, upon using Green's functions and recognizing integral forms, the recovery temperature field can be represented as

$$u_n(0,z,t) = u_{n-2}(0,z,t) - \frac{1}{\beta}\sqrt{\frac{\alpha}{\pi}} \times \int_{t_{n-2}}^{t_{n-1}} \frac{1}{\sqrt{(t-t')}} \frac{\partial u_{n-1}}{\partial z}(0,z,t')dt'. \tag{35}$$

Here, the recovery temperature field is an explicit integral involving the previous extraction phase instead of a differential equation. By substituting Eq. (34) into Eq. (35) and recognizing the integral form of first recovery shown in Eq. (17), the temperature for any recovery phase is

$$u_n(0,z,t) = u_2(0,z,t^{(n-1)};\Delta t_n) + u_{n-2}(0,z,t) + \frac{\beta}{\sqrt{\pi\alpha}} \int_0^{\Delta t_n} \frac{1}{\sqrt{t^{(n-1)} - t^{(n-2)}}} \times \left[\frac{\partial u_{n-2}}{\partial z}(0,z,t^{(n-1)} + t_{n-1}) + \frac{\partial F_n}{\partial z}(0,z,t^{(n-1)})\right] dt^{(n-2)} \tag{36}$$

Where

$$F_n(z, t^{(n)}) = \frac{\beta}{\sqrt{\pi\alpha}} \int_0^{t^{(n)}} \times \int_0^z \left(\xi^2 - \frac{1}{2}\right) \frac{\exp(-\xi^2)}{\tau^{3/2}} u_{n-2}(0, z', t-\tau)dz'd\tau, \tag{37}$$

Where t_n represents the duration of recovery and is equal to $t_n - t_{n-1}$.

The expression for the temperature along the fracture for subsequent recovery/extraction phases indicates that the temperature during extraction or recovery is dependent on a delayed fresh extraction or recovery plus the continuation of the previous recovery phase and finally a convolution integral involving the previous recovery phase. Although Eq. (36) gives us a more physical understanding, Eq. (35) is easier to implement numerically.

Multifracture EGS Reservoir

To extend the analysis to a multifracture reservoir, one can imagine an infinite series of rectangular fractures evenly spaced by a distance x_s. Because of the resulting symmetry, the heat flux at $x = x_s/2$ (the midplane between two fractures) is equal to zero. Under this formulation, the boundary condition given by Eq. (4) is modified to $\partial T/\partial x = 0$ at $x = x_s/2$ and the temperature field can be solved via Laplace transforms. The solution for a multifracture system in Laplace space is

$$\bar{u}(x, z, s) = \frac{u_0}{s} \exp\left[-r\beta \tanh\left(\frac{1}{2}rx_s\right) z\right] \left\{\cosh(rx) + \tanh\left(\frac{1}{2}rx_s\right)\sinh(rx)\right\}, \tag{38}$$

Where $r = \sqrt{s/\alpha}$. For the temperature at the outlet, the result becomes

$$\tilde{u}(0, z, s) = \frac{u_0}{s} \exp\left[-r\beta \tanh\left(\frac{1}{2} r x_s\right) z\right]. \tag{39}$$

As the spacing between fractures, x_s, goes to infinity, the tanh term goes to 1, as expected, and the resulting temperature field resembles that of a single fracture, given by Eq. (8). Because $_x$ determines the order of the extent of heat transfer that has taken place in the reservoir, the single fracture solution is valid when $_x/x_s \ll 1$. We can conjecture that when $_x/x_s \sim 1$, the solution for the single fracture should begin to deviate from the real solution as the fractures start to thermally interact with each other. Since an analytical inverse Laplace transform to Eq. (39) does not exist, we will rely on the numerical solution described below to characterize the conditions under which the heat extracted from a fracture is affected by interference by neighboring fractures.

TOUGH2 NUMERICAL MODEL

The TOUGH2 reservoir code developed at the Lawrence Berkeley National Laboratory (Pruess et al., 1999) was used for the numerical simulations of the extraction and recovery operational phases. TOUGH2 is a general purpose numerical simulation program for multidimensional non-isothermal flows of multiphase, multicomponent fluid mixtures in fractured and porous media. It employs an integral finite difference method (IFDM) in discretizing the medium, which has the advantage of irregular discretization in multiple dimensions. Time is discretized fully implicitly as a first-order backward finite difference and fluxes are computed using upstream weighting (Pruess et al., 1999). Please refer to Appendix B for the specifics of the numerical model, including geometry and mesh. Fig. 2 shows the implemented mesh and Fig. 3shows the simulation space.

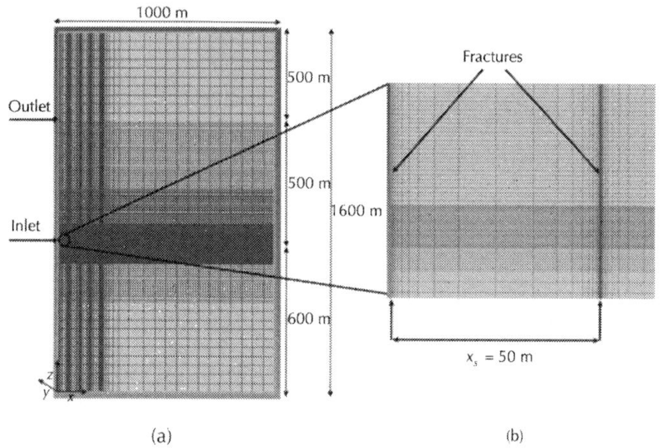

Figure 2: TOUGH2 mesh used for single and multifracture simulations. (a) Overview of the mesh with dimensions and inlet and outlet position identified. (b) Close-up to show the fractures in the mesh.

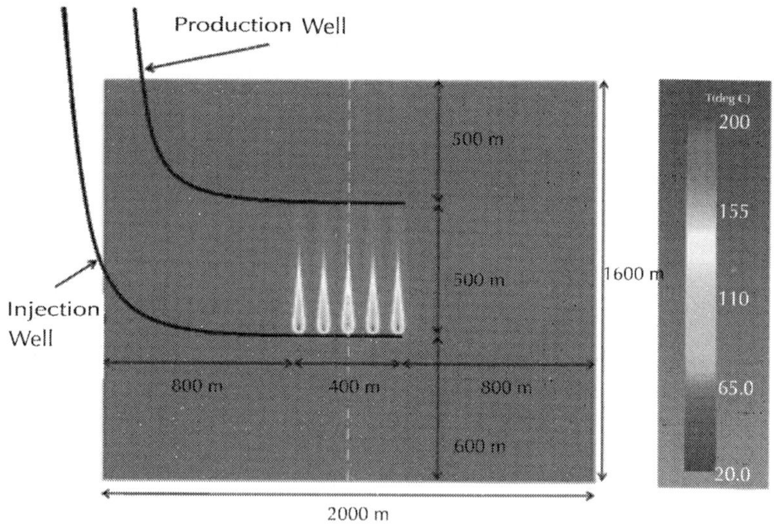

Figure 3: Dimensions of the simulation space used in TOUGH2. The dashed line down the middle indicates symmetry and thus only one side of the reservoir needs to be simulated. The presented reservoir has 5 fractures and has already undergone 10 years of extraction.

RESULTS AND DISCUSSIONS

Single Fracture Simulations

The single fracture simulations were first run with a flow rate of m/A=1.6*10^{-5} kg/ (m^2 s). The flow rate was chosen so that the thermal front remains within the reservoir so that the temperature varies within the fracture during much of the extraction phase. Fig. 4 compares the numerical and analytical results for the normalized temperature Θ during a 30 year extraction period at 4 different positions along the fracture. Note that the simulated fluid pressure does not vary more than 10% from the initial condition, as intended in the problem formulation. In general, the numerical results show somewhat less thermal drawdown than the analytical model with the largest deviation occurring at positions near the inlet. The deviation is a result of two-dimensional conduction in TOUGH2. Dimensional effects on the rate of thermal drawdown have also been observed by Kolditz (1995) and Cheng et al. (2001) in which increased dimensions led to less drawdown.

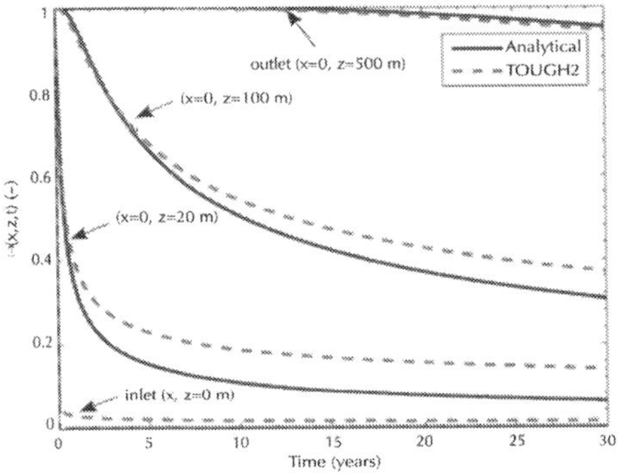

Figure 4: First extraction TOUGH2 numerical results compared to Eq. (6) at four different positions along the fracture for a 30 year extraction period and a mass loading of m/A=1.6*10^{-5} kg/(m^2 s). The black dashed line represents the mesh convergence fit as z 0 (refer to Appendix B).

Note that the analytical solutions only consider one-dimensional heat conduction and the error as a result of two-dimensional conduction scales as β^2. Using a flow rate that is three times higher, m/A=4.8*10^{-5} kg/ (m^2 s), β^2 is reduced by a factor of nine and the deviation between the analytical and numerical results should be minimized. Fig. 5 demonstrates the effect of β^2 and confirms the expectation based on the scaling analysis. As before, the temperature from the simulation results are higher in the simulations. The remaining two-dimensional conduction at small β may result primarily from conduction from the bulk rock below the fracture. Indeed, we found that setting the thermal conductivity of a row of grid blocks below the injection points to zero caused the simulation results to nearly coincide with the analytical solution. The agreement between the simulation and analytical results signifies that in a real reservoir, heat conduction parallel to the fracture is negligible compared to the heat transfer perpendicular to the fracture.

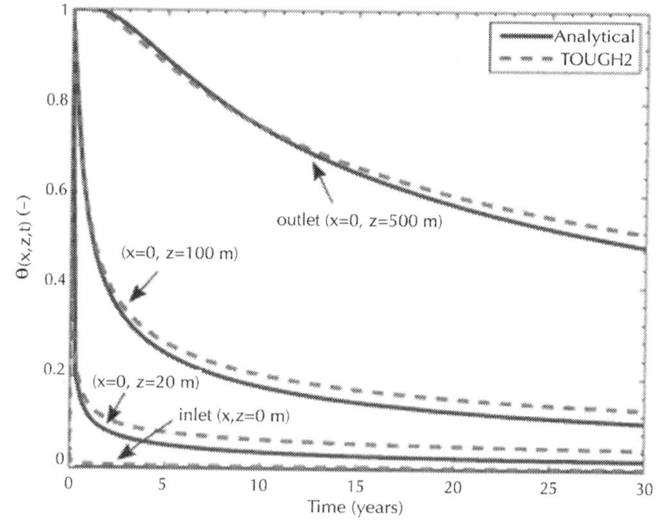

Figure 5: First extraction TOUGH2 numerical results compared to Eq. (6) at four different positions along the fracture for a 30 year extraction period and an area normalized flow rate of m/A=4.8*10^{-5} kg/(m^2 s).

Fig. 6 shows the first recovery simulation results compared with those obtained from Eq. (19). As expected, the bulk rock above the outlet has enhanced recovery. During extraction, forced convection in the fracture kept the ratio of conduction in z to conduction in x small. The convective heat transport rate along the fracture was balanced by the conductive heat transfer rate from the rock into the fracture in the x direction. During recovery, the forced convection is absent. Conduction in z would be expected to have less of an effect in the middle of the fracture where the variation of temperature with z is smaller than near the inlet and outlet and indeed the deviation of the numerical results from the analytical model is smaller at intermediate z in Fig. 6. As was discussed previously with respect to the analytical equation for recovery, the reservoir does not recover completely. The bulk of recovery occurs within a time period equal to that of the preceding extraction. After this time period, the deviation of the temperature from the bulk rock decays with a slow $1/\sqrt{t}$ behavior characteristic of an impulsive line sink of heat. As a result, the additional gains from allowing the reservoir to recover for time periods greater than about 3 to 5 times the extraction time are small.

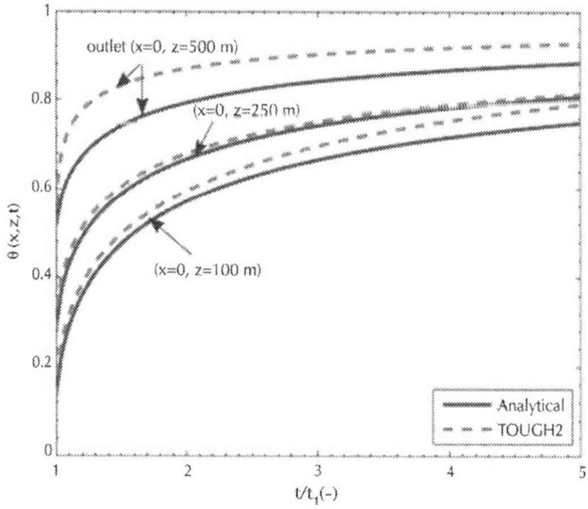

Figure 6: Comparison of first recovery results from TOUGH2 and Eq. (19) (with 100 terms) at three different positions along the fracture. The recov-

ery occurred after 30 years of extraction with an area normalized flow rate of $m/A=4.8*10^{-5}$ kg/ (m^2 s). Note that recovery starts at t_1 where t/t_1 = 1 represents the start of recovery.

The purpose of letting a reservoir recover is to be able to use the reservoir for extraction again with a similar performance as obtained in the first extraction. Fig. 7 validates the accuracy of the second extraction solution provided by Eq. (30). As discussed earlier, the added bulk rock had the effect of enhancing recovery near the outlet. However, these gains are quickly lost during a renewed extraction phase. Once the extraction starts, the temperature at the outlet is controlled by the heat convected from the middle of the fracture whose recovery is accurately predicted by the analytical model. According to Fig. 7, the difference in the outlet temperature between second extraction and the delayed fresh extraction is decreasing with time. The same conclusion is reached by looking at Eq. (30) where the second extraction is dominated by the delayed fresh extraction at larger times when both the recovery and convolution terms have decayed.

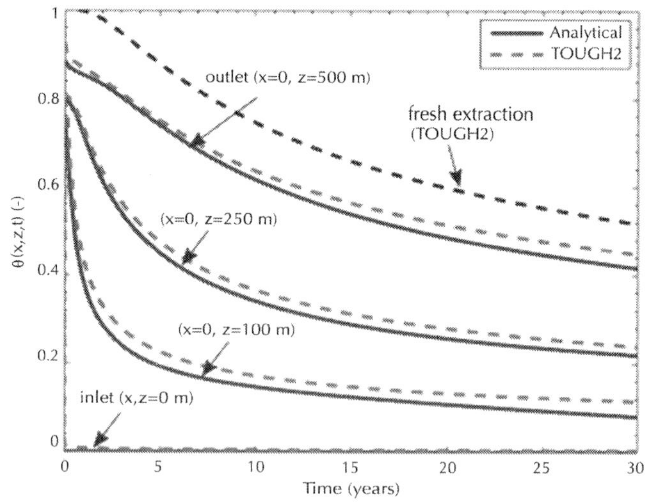

Figure 7: Second extraction TOUGH2 numerical results compared to Eq. (30) at four different positions along the fracture and for a 30 year extrac-

tion period and an area normalized flow rate of m/A=4.8*10^{-5} kg/ (m^2 s). The black dashed curve represents the outlet temperature of a fresh reservoir undergoing its first extraction phase.

Fig. 8 shows a similar behavior for the second recovery as that observed during the first recovery phase. The bulk of the recovery occurs early on and deviation between the analytical and numerical simulation results are small except near the ends of the fracture.

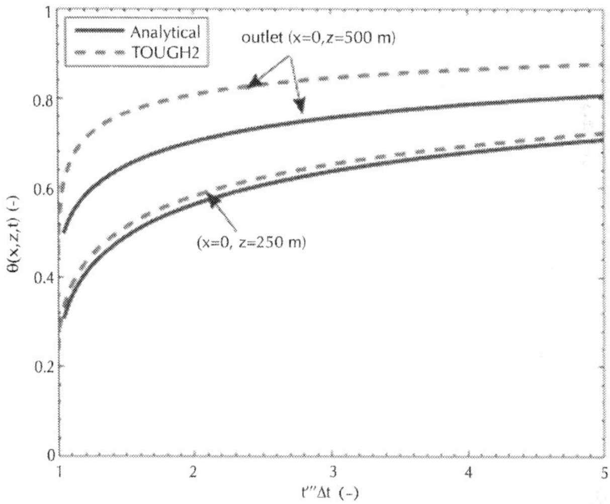

Figure 8: Comparison of second recovery results from TOUGH2 and Eq. (35) at two different positions along the fracture. The recovery occurred after a 30 year second extraction period with an area normalized flow rate of m/A=4.8*10^{-5} kg/ (m^2 s). Note that $t\Delta = t - t_2$ and second extraction starts at t_2 while second recovery starts at t_3 with $\Delta t = t_3 - t_2$ (the duration of the second extraction phase). $t/\Delta t = 1$ represents the start of second recovery.

Multiple Fractures Simulations

A multiple parallel fractured reservoir is a more realistic representation of an EGS reservoir than a single fracture reservoir (Armstead and Tester, 1987). Multiple fractures have the effect of spreading out the mass loading into the reservoir and thus increasing the effective heat transfer area. Thus, one would expect longer production lifetime for a multifracture EGS reservoir.

In these simulations the total flow rate in TOUGH2 was set to 70 kg/s corresponding to a total mass loading of 1.4×10^{-4} kg/(m^2 s) for a fracture system with a one side area of 500,000 m^2 (500 m long and 1000 m wide). Economic total flow rates for EGS are estimated to range from 50 to 150 kg/s for a single reservoir (Tester et al., 2006). Note we are modeling only half of the fracture system due to symmetry. In this case, each fracture receives $0.14/n$ kg/s where n is the number of fractures accounting for symmetry. The extraction phase was set to 10 years while the recovery phase was set to 20 years.

Fig. 9 shows the results of the TOUGH2 multifracture simulations. The temperature at the outlet of the middle fractures and the outer fractures within any given reservoir does not vary more than one percent. This indicates that the temperature behaves in a manner similar to the infinite array of fractures used in formulating Eq. (39).

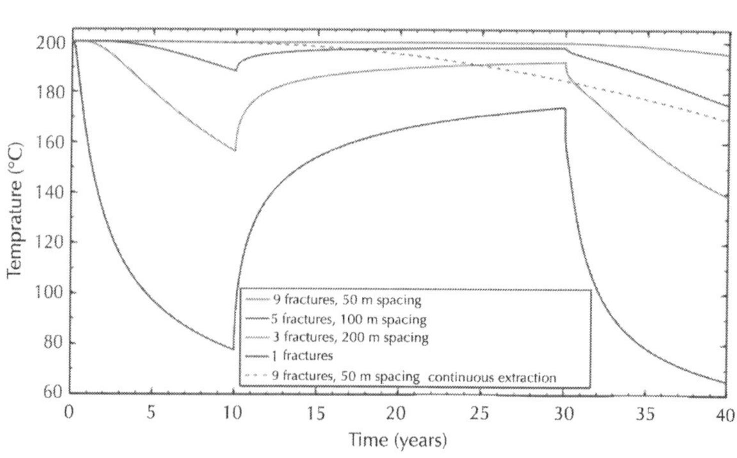

Figure 9: Average of the outlet temperatures of the fractures during 10 years of extraction and 20 years of recovery. Because of a low drawdown rate for the 9 fracture reservoir, a 40 year continuous extraction was plotted.

As expected, the decrease in the mass loading per fracture as one increases the number of fractures slows the rate of drawdown. In going from one ($x_s = \infty$) to three fractures ($x_s = 200$ m) the outlet temperature after 10 years of extraction increases from 80 °C to about 160 °C. With 5 ($x_s = 100$ m) and 9 fractures ($x_s = 50$ m), the drawdown remains within acceptable levels, from the point of view of power conversion, even at the end of the first extraction phase. A 20-year recovery phase is able to raise the outlet temperature sufficiently to begin a new extraction phase. However, as might be expected, the second extraction suffers from a more rapid drawdown, putting a burden on the surface installation to adapt more quickly to lower temperatures. Nonetheless, the recovery curves indicate that there would be no substantial gain for extending recovery beyond 30 years and that the ratio of recovery time to extraction time should fall between 2 and 4.

Another way to interpret the production results is to look at the total thermal energy content (enthalpy) and the maximum electrical work (exergy) of the fluid produced from the reservoir. Exergy or availability is the maximum possible work obtained from bringing a heat source to equilibrium with the surroundings at a specified temperature and pressure "dead state". Because exergy relates to the maximum work, it is the appropriate thermodynamic quantity for power generation while for direct use, enthalpy is more relevant. Fig. 10 shows the enthalpy extracted as a percentage of the enthalpy (above the inlet water temperature) available in the reservoir region, defined to be 400 m × 500 m × 1000 m (0.20 km^3) volume. The exergy produced is expressed as a percentage of that which would be produced if the produced water reached the bulk rock temperature. The dead state was defined as 20 °C and 20 bar. Fig. 10 reveals that for the given flow rate and fracture distribution investigated, the amount of enthalpy and exergy extracted for 5 fractures is nearly the same as in the case of 9 fractures. For the 5 and 9 fracture reservoir, breakthrough has yet to reach the outlet and thus there is no benefit with regards to outlet temperature to having more fractures beyond 5. In such a situation, an increase in m/A would differentiate the amounts of extracted enthalpy and exergy for 5 and 9 fractures. Comparison of the single fracture and 3 fracture reservoirs shows that the increase in the total extracted enthalpy per fracture has a trade-off in terms of decreased quality of extracted heat. A threefold increase in the mass loading per fracture for the single fracture reservoir as compared to the 3 fracture model resulted in considerable drop in the specific exergy of the fluid. Proper management of the EGS reservoir will require balancing the total extracted enthalpy and the specific exergy of the produced fluid with respect to anticipated production lifetime of the reservoir. This balance would be achieved by adjusting the mass flow into the reservoir depending on its size and the surface infrastructure operation.

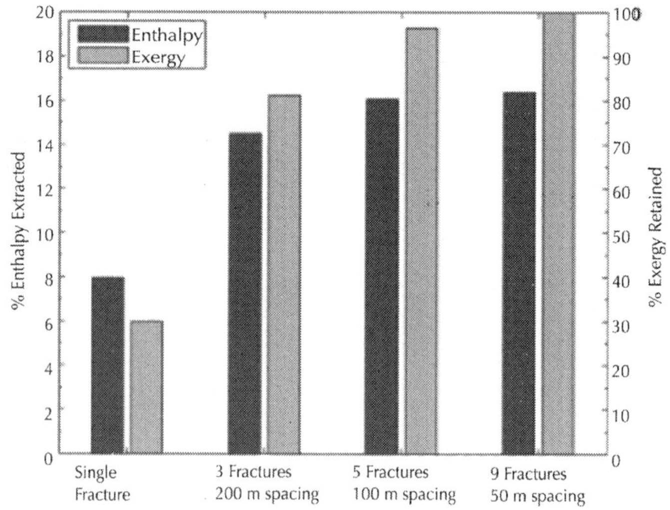

Figure 10: Percent of enthalpy extracted and exergy retained after 10 years of extraction by a total $m/A=1.4*10^{-4}$ kg/ (m^2 s) from an EGS reservoir normalized by the initial enthalpy of the entire reservoir. The reservoir is defined to be 400 m × 500 m × 1000 m (0.20 km^3) and the reference temperature and pressure was defined as 20 °C and 20 bar.

Fig. 11 is a plot of the multifracture analytical solution presented in Eq. (39) along with the results from the multifracture simulations. In the figure, the results indicate a close agreement between Eq.

(39) and the numerical results. The small deviation is caused by two-dimensional heat conduction. From the numerical model, the temperature of each fracture was nearly identical. In particular, the outer fracture with one neighbor had a similar temperature profile to the middle fracture with two neighbors. The uniformity among the temperature profiles within the fractures suggests that the fractures are not thermally interfering with one another. Interestingly, the analytical solution for a single fracture outlet temperature with the same m/A as each fracture in a multifracture EGS reservoir results in a set of curves indistinguishable from those derived from Eq. (39). Because the two sets of curves are effectively identical, there is no measurable thermal interaction for the entire set of fracture spacings and flow rates investigated in this study as noted earlier. Lack of thermal interaction is seen in Fig. 3 where the thermal depletion zone for each fracture is clearly distinguishable and has yet to overlap. The corresponding values of δ_x/x_s with $\delta_x = 2\sqrt{\alpha t}$ for 50 m, 100 m, and 200 m are 0.71, 0.36, and 0.18 respectively after 10 years of extraction. Although δ_x/x_s for fracture spacings of 50 m and 100 m are of order 1, the outlet temperature has not been effected by the thermal interaction even though $\delta_x \sim x_s$. As δ_x gets closer to x_s, temperatures will first start to deviate from the single fracture case in the injection region ($z \approx 0$) but do not propagate to the outlet. Additionally, when thermal penetration in the x direction at points near the inlet region starts to interfere with the thermal disturbance of the adjacent fracture, the thermal disturbance near the outlet is much smaller than the fracture spacing and the temperature field behaves as if the fractures were infinitely separated. Thus, points near the outlet are not affected by the thermal disturbance caused by the adjacent fracture.

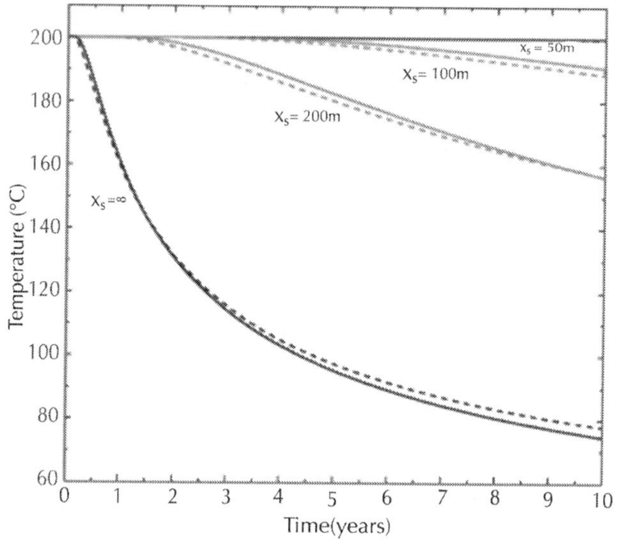

Figure 11: Outlet temperature results obtained from a parallel multifracture reservoir modeled in TOUGH2 (dashed curves) compared to multifracture solution (solid curves) from Eq. (39).

For fracture spacings of 50 m, 100 m, and 200 m, a 5% production temperature deviation with respect to the single fracture solution occurs after 35, 46, and 120 years, respectively. These times correspond to δ_x/x_s values of 1.3, 0.76, and 0.62, respectively, all of order 1. Thermal interaction is not exclusively a function of δ_x/x_s but depends on β as well. For the cases of 1, 3, 5, and 9 fractures, the β values are 0.0020, 0.0059, 0.099, and 0.18, respectively. These results indicate that for systems with sufficient fracture spacings, parallel fractures do not interact thermally and one can model them using a single fracture model with β or m/A based on the flow rate passing through each fracture.

We can define an efficiency that describes how effective recovery was in restoring the farmed thermal energy of the reservoir and effecting the success of a second extraction phase. The efficiency η can be defined with either enthalpy or exergy. It is important to be able to predict how much less energy one will extract due to an incomplete recovery and make the appropriate surface installation adjustments. The efficiency of the second extraction is the ratio of the enthalpy or exergy extracted during the second extraction phase to that extracted during the first extraction phase. The efficiency was calculated using the analytical equations for a single fracture for flows corresponding to the 1, 3, and 5 fracture reservoirs. Fig. 12 and Fig. 13 depict the two differently defined efficiencies under 5, 10, 15, and 20 year extraction periods as a function of the ratio of recovery to extraction time.

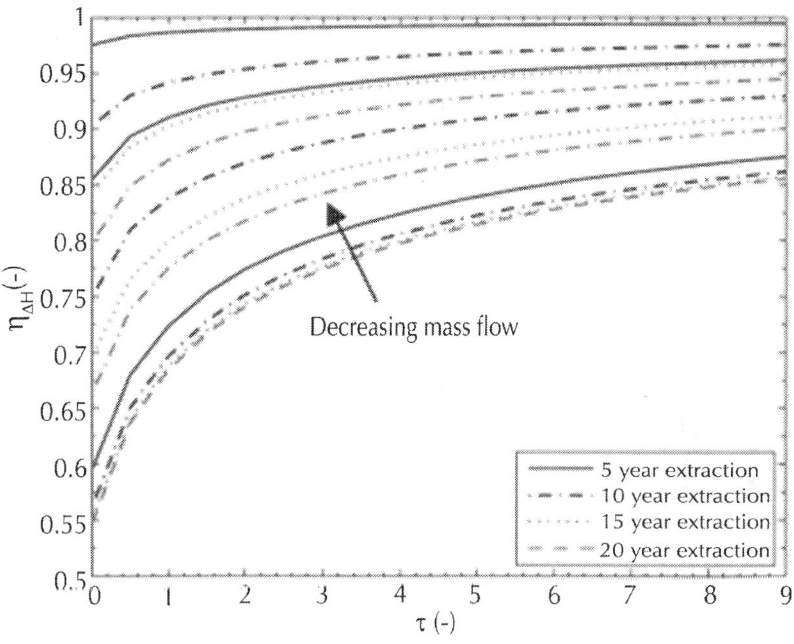

Figure 12: Second extraction efficiency in terms of enthalpy for a 5, 10, 15, and 20 year extraction period as a function of the duration of recovery to extraction, τ. A τ value of zero represents a recovery duration of zero while one represents equal recovery and extraction durations. Because colors are repeated, the first instance of a particular color, from the bottom to the top of the figure, represents the single fracture case while the second and third represent the 3 and 5 fracture reservoir, respectively. (For interpretation of the references to color in this figure legend, the reader is referred to the web version of the article.)

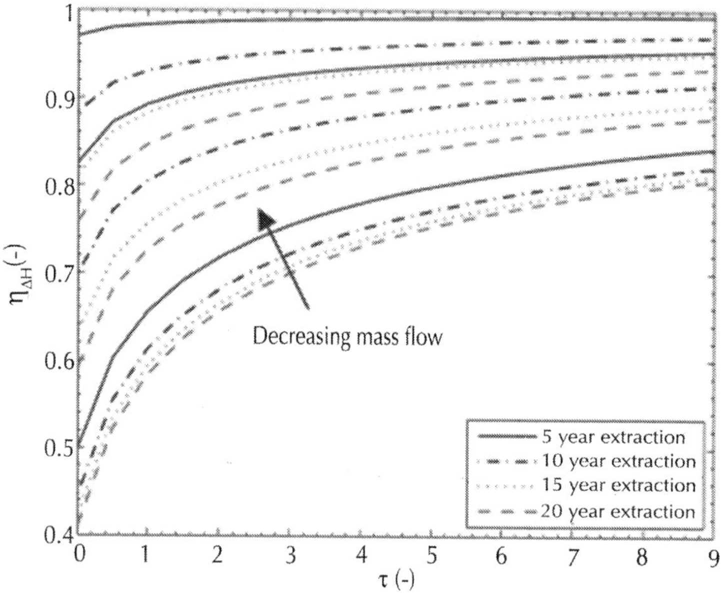

Figure 13: Second extraction efficiency in terms of exergy for a 5, 10, 15, and 20 year extraction period as a function of the duration of recovery to extraction, τ. A τ value of zero represents a recovery duration of zero while one represents equal recovery and extraction durations. Because colors are repeated, the first instance of a particular color, from the bottom to the top of the figure, represents the single fracture case while the second and third represent the 3 and 5 fracture reservoir, respectively. (For interpretation of the references to color in this figure legend, the reader is referred to the web version of the article.)

The results demonstrate that the second extraction efficiency is a function not only of the ratio of recovery to extraction time but also the mass flow into the reservoir. As was seen in the graphs of temperature during recovery and second extraction, Fig. 12 and Fig. 13 also suggest that little is gained for letting a reservoir recover beyond a ratio of recovery to extraction time of 4.

At higher flow rates or fewer fractures, the different extraction duration curves are grouped closer, revealing that there will be only a small increase of efficiency by reducing the extraction period. At smaller flow rates or more fractures, the grouping is much larger and shorter extraction periods yield second extraction rates closer to the first extraction period. Under these conditions, shorter extraction periods are favored because they result in higher second extraction efficiencies.

CONCLUSIONS AND RECOMMENDATIONS

The ability of conductively dominated EGS reservoirs to recover their thermal energy after ceasing production was assessed using both analytical and numerical methods. The assessment is important to demonstrate how Enhanced or Engineered Geothermal Systems could sustainably supply energy over a range of scales from a specific reservoir to a set of reservoirs operating sequentially in a field. With such modular systems operating successfully, one can envision how EGS could provide a significant portion of our global energy needs in a sustainable manner. This important characteristic provides a rationale for further field testing of EGS under commercial scale production conditions.

Analytical equations were derived for the temperature along a single rectangular fracture of an EGS reservoir during alternating periods of thermal extraction and recovery using a Green's function solution of the transient one-dimensional heat conduction in the rock. The analytical equations were shown to agree well with a numerical solution obtained using TOUGH2 of the energy conservation including two-dimensional conduction. The analytical equations can therefore be used as an alternative to a numerical model and as a zeroth order model to explore a particular facet of EGS operations such as thermal recovery.

The majority of the equations derived here are for a single fracture, but under certain conditions, the single fracture model can be used to model a system of parallel fractures, specifically, under typical fracture spacings and time scales of extraction. If the fracture spacing is large enough then the fractures do not thermally interact and the fractures behave independently. The single fracture model can be used for multiple parallel fractures as long the mass flow or is scaled on a per fracture basis. With multiple fractures present, the rate of drawdown is significantly reduced. The slower rate of drawdown and the recoverability of the reservoir enhance the renewability and sustainability of conduction dominated EGS reservoirs.

Although conduction dominated EGS reservoirs do not have recovery times as small as those with substantial convective or advective drive, this study has shown that even these EGS reservoirs can be operated sustainably. A significant amount of their original thermal energy will be restored following a defined production period if they are allowed to recovery for two to four times the initial extraction time. The time scales of recovery are within the life time of surface installations which could allow for the recycling of an old reservoir. This study affirms classification of EGS reservoirs as renewable and capable of sustainable energy production.

The investigated models can be considered a conservative representation of a worst case because they do not account for recharge by convection. Even though a low permeability EGS reservoir may not initially contain measurable amounts of fluid, water that had been introduced into the fracture system will lead to convective or advective flows. Fluid loss to the far field occurs because rock outside the fracture zone has a finite permeability and porosity. Because reservoir fluid pressures are expected to exceed the *in situ* pore pressures in EGS reservoirs with injection pumps, the injected water will have a tendency to migrate out of the fractures zones into the rock matrix during extraction. During recovery or after reducing production pressures, the "lost" fluid will flow back into the active reservoir region and accelerate recovery via the addition of convectively driven hot water. Additionally, since cooler regions of the reservoir will exhibit larger apertures caused by thermal contraction of the rock, these regions will receive the majority of the flow and expedite recovery in places where it is needed the most.

Along with neglecting convection, the idealized approach used in this study did not consider non-parallel or intersecting fractures or buoyancy driven flow, which are important features in real EGS reservoirs (Harlow and Pracht, 1972, McFarland and Murphy, 1976 and Armstead and Tester, 1987). Furthermore, as was discussed in the introduction, our model does not include some of the coupled thermal, hydraulic, mechanical, and chemical physics that will surely effect how the system behaves. These features could either increase or reduce extraction and/or recovery times depending on their extent. The effect these features have on the rate of recovery is unknown and should be investigated in the future.

The optimal operational strategy of an EGS reservoir will vary with the end-use of the surface installation. For example, a district heating system can allow for somewhat greater thermal drawdown compared to a binary power plant which is more sensitive to temperature changes in the production fluid. Additionally, the best strategy may involve different cycle times and variable flow rates and include tandem use of reservoirs. The models developed in this study can be used as a basis for exploring the optimal strategies for using EGS reservoirs for specific applications.

ACKNOWLEDGMENTS

The authors would like to thank the Department of Energy and the National Science Foundation through the IGERT program for partial funding.

APPENDIX A. DERIVATION OF THE INFINITE SERIES RECOVERY SOLUTION

The variable transformation *u* results in homogeneous boundary and initial conditions and simplifies obtaining a solution. The transformations of Eq. (14) and its boundary and initial conditions are

$$\frac{\partial u}{\partial t} - \alpha \frac{\partial^2 u}{\partial t^2} = \frac{q(z,t)\delta(x)}{\rho_r c_{p,r}}, \tag{A1}$$

$$\lim_{x \to \pm\infty} u(x,z,t) = 0, \tag{A2}$$

And

$$u(x,z,t \leq 0) = 0. \tag{A3}$$

The appropriate Green's function for the domain of $x \in R$ and $t \in R_0^+$ is shown in Eq. (16). Following the common approach for applying Green's functions, the solution to the PDE in Eq. (A.1) is the integral

$$u(x,z,t) = \int_0^{\infty} \int_{-\infty}^{\infty} G(x-x', t-t') \frac{q(x',z,t')\delta(x')}{\rho_r c_{p,r}} dx'dt'. \tag{A4}$$

The dx' integral can be evaluated by applying the sifting property of the Dirac delta function and the upper integration boundary for t' can be reduced from $t' = \infty$ to $t = t$ because of the step function. The result is

$$u(x,z,t) = \int_0^{t} \frac{1}{\sqrt{4\pi\alpha(t-t')}} \exp\left[-\frac{x^2}{4\alpha(t-t')}\right] \frac{q(z,t')}{\rho_r c_{p,r}} dt'. \tag{A5}$$

Because the heat flux at the rock/water interface is piecewise continuous with respect to time, as seen in Eq. (15), Eq. (A.5) is best broken into two components, an integral from 0 to t_1 and a second integral from t_1 to t. The integral from t_1 to t is zero because there is no heat flux at the rock/water interface during recovery. The heat flux $q\ (z, t)$ during the extraction phase can be determined by taking the derivative of the temperature profile for the extraction in Eq. (6). The result is

$$q(z,t) = \frac{2k_r u_0}{\sqrt{\pi\alpha t}} \exp\left[-\frac{\beta^2 z^2}{4\alpha t}\right], \quad 0 < t \leq t_1. \tag{A6}$$

Substituting the heat flux shown in Eq. (A.6) into Eq. (A.5), we get the integral solution for recovery,

$$u(x,z,t) = \frac{u_0}{\pi}\int_0^{t_1} \frac{1}{\sqrt{t'}\sqrt{t-t'}} \exp\left[-\frac{1}{4\alpha}\left\{\frac{x^2}{(t-t')}+\frac{\beta^2 z^2}{t'}\right\}\right] dt'. \quad \text{(A7)}$$

Eq. (A.7) can be rewritten as

$$\frac{u(0,z,t)}{u_0} = \frac{1}{\pi}\int_0^{t_1} \frac{1}{\sqrt{t'}\sqrt{t}} \frac{1}{\sqrt{1-(t'/t)}} \exp\left[-\frac{\beta^2 z^2}{4\alpha t'}\right] dt'. \quad \text{(A8)}$$

Using the Taylor series expansion for the term $\sqrt{1-t'/t}$ about 0, Eq. (A.8) can be simplified to

$$\frac{u(0,z,t)}{u_0} = \frac{1}{\pi}\int_0^{t_1} \frac{1}{t^{n+(1/2)}} \sum_{n=0}^{\infty} \binom{n-(1/2)}{n} t'^{n-(1/2)}$$

$$\exp\left[-\frac{\beta^2 z^2}{4\alpha t'}\right] dt'. \quad \text{(A9)}$$

The integral in each term of the summation in Eq. (A.9) can be expressed as an incomplete gamma function,

$$\frac{u(0,z,t)}{u_0} = \frac{1}{\pi}\sum_{n=0}^{\infty} \binom{n-(1/2)}{n} \left(\frac{t_1}{t}\right)^{n+(1/2)} \xi_r^{2n+1}(x,z,t_1)$$

$$\Gamma\left(-\left\{n+\frac{1}{2}\right\}, \xi_r^2\right) \quad \text{(A10)}$$

With the incomplete gamma function defined as

$$\Gamma(a,s) = \int_s^{\infty} \tau^{a-1} e^{-\tau} d\tau. \quad \text{(A11)}$$

Applying the substitution $\tau = x^2$, the incomplete gamma function can be evaluated using the following integral found in Gradshteyn and Ryzhik (2007)

$$\int_u^\infty \frac{e^{-p^2x^2}}{x^{2n}}dx = \frac{(-1)^n 2^{n-1} p^{2n-1}\sqrt{\pi}}{(2n-1)!!}\operatorname{erfc}(pu)$$

$$+\frac{e^{-p^2u^2}}{2u^{2n-1}}\sum_{k=0}^{n-1}\frac{(-1)^k 2^{k+1}(pu)^{2k}}{(2n-1)(2n-3)\cdots(2n-2k-1)}, \quad p>0 \tag{A12}$$

And substituted in Eq. (A.10) to yield the infinite series solution given in Eq. (19).

APPENDIX B. NUMERICAL MODEL DESCRIPTION

The computational domain in TOUGH2 was 1000 m in the *x* direction, 1600 m in the *z* direction, and 1 m in the *y* direction. Identical to the analytical solution, there is no heat or mass flow in the *y* direction and as a result, the dimension in *y* is arbitrary. Only half of the reservoir needs to be simulated because of symmetry along the *yz* plane. As a result, the total flow rate used in the simulation needs to be scaled down by a factor of 2*H*. Hence, for our reservoir geometry, to represent a total flow of 70 kg/s into a fracture that is 1000 m long in *y*, a flow of 0.035 kg/s is used. Any confusion regarding the length of the *y* domain can be eliminated by using m/A where m refers to the total flow into the system spanning ±*x*. The mesh is adequately sized with respect to *x* and *z* as the thermal penetration length $_x$ in typically crystalline rocks with $\approx 10^{-6}$ m^2/s will only reach 1000 m after 10,000 years.

In order to simulate systems with single and multiple parallel fractures, a modular mesh was created. The mesh is called modular because by changing material properties of certain grid blocks, several different fracture scenarios can be simulated with only one mesh. The fractures are contained within a 200 m wide in *x* by 500 m long in *z* by 1 m in *y* simulation volume as seen in Fig. 3. The fractures are 0.06 m wide. Because the mesh needs to be refined in regions of higher temperature gradients, such as near the fracture, a periodic logarithmic mesh spacing in x was used. For every 50 m in x, there are 27 grid blocks between each fracture which logarithmically increase from 0.06 m and then logarithmically decrease at the midpoint so that the last grid block in the period before the next fracture is back at 0.06 m. Given the fracture spacing of 50 m, a total of 4.5 full fractures can be simulated. The half fracture results from the plane of symmetry cutting through the central fracture. Because of symmetry, the model can simulate nine full fractures in a reservoir. The mesh has 237 grid blocks in the z direction whose length varies as a step function. For example, near the inlet there are 40 0.250 m thick grid blocks followed by another 40 grid blocks but that are 1 m thick. Because heat and mass do not flow in the y direction, the mesh in y was not refined and only consists of one block. As seen in Fig. 3, bulk rock was added above the production and below the injection well of the fracture to account for heat conduction beyond the fracture length. The bulk rock above and below the production and injection well will have the most effect in the recovery phases because conduction in z brings heat from the additional rock mass in the system to the thermally depleted zone.

The mesh is primarily made up by two different materials named "rock" and "fracture". To model a scenario with fewer fractures, the material properties of the appropriate grid blocks that were designated as the fracture material are changed to the rock material. The permeability and porosity of the rock material was set to zero. Thus, no fluid can leave or enter the fracture grid blocks because the grid blocks adjacent to the fractures are designated as rock. The porosity of the fracture material was set to one to model the flow in an open fracture while the permeability was arbitrarily set to 10^{-9} m^2 to minimize the pressure drop. By keeping the pressure drop small with a relatively large permeability, single phase flow in the fracture will be ensured. The initial temperature and pressure are 200 °C and 20 bar.

The injection cell is represented by an extra cell superimposed over the first grid block of each fracture which supplies a specified source of fluid mass per unit time at the injection temperature. The production cell is located 500 m further down from each injection cell. The cell directly above the production cell in zacts as a sink of mass and has zero thermal conductivity. Any mass and convective heat flow that enters this cell is removed from the system.

To model the recovery phase, the mass flow was simply set to zero and the initial conditions were set to the final field value of the previous extraction phase. The values of the thermophysical properties of the rock in the simulations are ρ_r = 2700 kg/m^3, $c_{p,r}$ = 1050 J/(kg °C), k_r = 2.9 W/(m °C), α = 1.02 × 10^{-6} m^2/s. TOUGH2 uses an equation of state to determine the thermophysical properties of water and as a consequence, will vary with temperature and pressure. This is in contrast with the analytical equations which assume constant thermophysical properties. Nonetheless, the only property of water of importance in comparing numerical and analytical results is the heat capacity which varies from 4493 J/(kg °C) at 200 °C to 4178 J/(kg °C) at 20 °C, all at 20 bar.

During extraction, it is to be expected that the temperature of the rock adjacent to the inlet x, z = 0 will quickly be reduced to the inlet temperature so that Θ = 0. Fig. 4 shows, that the temperature simulated with TOUGH2 at x, z = 0 remains at a non-zero value for an extended period of time. We found that this behavior was due to the difficulty of resolving the rapidly varying temperature field in the vicinity of the inlet. In particular, the deviation of the numerical from the analytical solution was found to scale as $\sqrt{\Delta z}$ when the grid spacing Δz near the inlet was reduced. By applying a fit of the form $\Theta(\Delta z, t) = \Theta(0, t) + a\sqrt{\Delta z}$ for Δz = 0.250, 0.125, 0.0625 m, and extrapolating the value of Θ at the inlet for an infinitely small grid, Θ was found to be as small as 10^{-3} after less than a month. The extrapolated temperature profile at the inlet is plotted as the black line in Fig. 4. At higher flow rates such as the value m/A=4.8*10^{-5} kg/ (m^2 s) presented in Fig. 6, the convective heat transfer is strong enough so that the simulated inlet temperature decays to a very small value even without refining the grid.

REFERENCES

1. Armstead, H., Tester, J.W., 1987. Heat Mining, 1st ed. E. & F.N. Spon, London, New York.
2. Arpaci, V.S., 1966. Conduction Heat Transfer. Addison-Wesley, Reading.
3. Axelsson, G., Gudmundsson, A., Steingrímsson, B., Pálmason, G., Ármannsson, H., Tulinius, H., Flóvenz, O., Björnsson, S., Stefánsson, V., 2001. Sustainable production of geothermal energy: suggested definition. IGA-News Quarterly 43, 1–2.
4. Axelsson, G., Stefánsson, Björnsson, G., 2005. Sustainable utilization of geothermal resources for 100–300 years. Paper presented at Proceedings of the World Geothermal Congress, Antalya, Turkey.
5. Bromley, C., Mongillo, M., Rybach, L., 2006. Sustainable utilization strategies and promotion of beneficial environmental effects-having your cake and eating it too. Paper presented at New Zealand Geothermal Workshop.
6. Carslaw, H., Jaeger, J., 1959. Conduction of Heat in Solids, 2nd ed. Oxford University Press, Oxford.
7. Cheng, A.H.D., Ghassemi, A., Detournay, E., 2001. Integral equation solution of heat extraction from a fracture in hot dry rock. International Journal for Numerical and Analytical Methods in Geomechanics 25, 1327–1338.
8. Duffy, D.G., 2001. Green's Functions with Applications. Studies in Advanced Mathematics. Chapman & Hall/CRC, Washington, DC.
9. Elsworth, D., 1989. Theory of thermal recovery from a spherically stimulated hot dry rock reservoir. Journal of Geophysical Research 94, 1927–1934.
10. Ghassemi, A., Zhou, X., 2011. A three-dimensional thermo-poroelastic model for fracture response to injection/extraction in enhanced geothermal systems. Geothermics 40, 39–49.
11. Gradshteyn, I., Ryzhik, I., 2007. Table of Integrals, Series, and

Products, 7th ed. Academic Press, New York.

12. Gringarten, A.C., Witherspoon, P.A., Ohnishi, Y., 1975. Theory of heat extraction from fractured hot dry rock. Journal of Geophysical Research 80, 1120–1124.
13. Harlow, F.H., Pracht,W.E., 1972.Atheoretical study of geothermal energy extraction. Journal of Geophysical Research 77, 7038–7048.
14. Kohl, T., Evansi, K., Hopkirk, R., Rybach, L., 1995. Coupled hydraulic, thermal and mechanical considerations for the simulation of hot dry rock reservoirs. Geothermics 24, 345–359.
15. Kolditz, O., 1995. Modelling flow and heat transfer in fractured rocks: dimensional effect of matrix heat diffusion. Geothermics 24, 421–437.
16. McFarland, R.D., Murphy, H.D., 1976. Extracting Energy from HydraulicallyFractured Geothermal Reservoirs. ASME, State Line, NV.
17. Megel, T., Rybach, L., 2000. Production capacity and sustainability of geothermal doublets. Paper presented at Proceedings of the World Geothermal Congress, World Geothermal Congress, Kyushu, Tohoku, Japan, pp. 849–854.
18. Ogino, F., Yamamura, M., Fukuda, T., 1999. Heat transfer from hot dry rock to water flowing through a circular fracture. Geothermics 28, 21–44.
19. Pruess, K., Oldenburg, C., Moridis, G., 1999. TOUGH2 User's Guide, Version 2.0. Technical Report. Lawrence Berkeley National Laboratory.
20. Rybach, L., Megel, T., Eugster, W., 2000. At what time scale are geothermal resources renewable? Paper presented at Proceedings of the World Geothermal Congress, Kyushu, and Tohoku, Japan.
21. Sanyal, S.K., 2005. Sustainability and renewability of geothermal power capacity. Paper presented at Proceedings of the World Geothermal Congress, Antalya, Turkey.

22. Taron, J., Elsworth, D., 2009. Thermal-hydrologic-mechanical-chemical processes in the evolution of engineered geothermal reservoirs. International Journal of Rock Mechanics and Mining Sciences 46, 855–864.
23. Tester, J.W., Anderson, B.J., Batchelor, A.S., Blackwell, D.D., DiPippo, R., Drake, E.M., Garnish, J., Livesay, B., Moore, M.C., Nichols, K., Petty, S., Toksoz, M.N., Veatch, R.W., Baria, R., Augustine, C., Murphy, E., Negraru, P., Richards, M., 2006. The Future of Geothermal Energy. Technical Report DOE Contract DE-AC07- 05ID14517. Massachusetts Institute of Technology.
24. Willis-Richards, J., Wallroth, T., 1995. Approaches to the modelling of HDR reservoirs: a review. Geothermics 24, 307–332.
25. Wunder, R., Murphy, H., 1978. Thermal drawdown and recovery of singly and multiply fractured hot dry rock reservoirs. Informal Report W-7405-ENG. 36. Los Alamos Scientific Laboratory of the University of California.

Chapter 2

A Reservoir Simulation Approach for Modeling of Naturally Fractured Reservoirs

H. Mohammadi[a], A. Khaksar Manshad[a], and G.H. Montazeri[b]

[a]Department of Chemical Engineering, School of Engineering, Persian Gulf University, Bushehr 75169, Iran

[b]Department of Chemical and Petroleum Engineering, Mahshahr Branch, Islamic Azad University, Mahshahr, Iran

ABSTRACT

In this investigation, the Warren and Root model proposed for the simulation of naturally fractured reservoir was improved. A reservoir simulation approach was used to develop a 2D model of a synthetic oil reservoir. Main rock properties of each gridblock were defined for two different types of gridblocks called matrix and fracture gridblocks. These two gridblocks were different in porosity and permeability values which were higher for fracture gridblocks compared to the matrix gridblocks. This model was solved using the implicit finite difference method. Results showed an improvement in the Warren and Root model especially in region 2 of the semilog plot of pressure drop versus time, which indicated a linear transition zone with no inflection point as predicted by other investigators. Effects of fracture spacing, fracture permeability, fracture porosity, matrix permeability and matrix porosity on the behavior of a typical naturally fractured reservoir were also presented.

INTRODUCTION

Naturally fractured reservoirs (NFRs) have been studied intensively for several decades in the geologic and engineering fields and recently, two new approaches have been proposed to address the problem of transient pressure testing in naturally fractured reservoirs: using fractals to describe the reservoir geometry, and using numerical simulation to model fluid flow in a fracture network embedded in a matrix [8].

Based on the theory of fluid flow in naturally fractured porous media developed in the 1960s by Barenblatt and Zheltov [1], Warren and Root [17] introduced the concept of dualporosity models into petroleum reservoir engineering. Their idealized model consisting of a highly interconnected set of fractures, which is supplied by fluids from numerous small matrix blocks, is shown in Fig. 1.

Whereas the matrix permeability is much smaller than the fracture permeability, the fracture porosity of a particular class of NFRs seldom exceeds 1.5% or 2%, and usually falls below 1% [12]. The high permeability of a fracture results in a high diffusivity of the pressure propagation pulse along the fracture; however, due to a significant contrast between matrix and fracture permeabilities, the matrix has a "delayed" response to pressure changes that occur in the surrounding fractures. In theory, the double-porosity behavior yields two parallel straight lines on a semilog plot, whereas there are neither wellbore nor outer boundary effects. The semilog plot consists of three sections: (i) the first straight line, which represents the homogeneous behavior of the naturally fractured medium before the matrix medium starts to respond (transient radial flow), (ii) a transition section (between two straight lines), which corresponds to the onset of interporosity flow; and (iii) the second semilog straight line, which represents the homogeneous behavior of composite media (fracture permeability with the sum of matrix and fracture storages) when recharge from the matrix medium is fully established. Also a fourth region may exist in the semilog plot of a typical NFR as shown in this work which is called the reservoir boundary effect.

The nature of matrix and fracture interaction is manifested during this transitional period of matrix-to-fracture fluid transfer by Streltsova [16]. The important features of the most common models are listed in Table 1.

At first a pseudo-steady state interporosity flow concept was proposed by Barenblatt and Zheltov [1] and in subsequent studies [17,10,5,7]. Various extensions of the model have been made by assuming transient interporosity flow [4,9,13,16,6,3,11].

To model fracture–matrix interaction, researchers have developed and applied many different conceptual models and modeling approaches as summarized by Berkowitz [2]. Commonly used mathematical methods include: (1) an explicit discrete-fracture and matrix model (e.g. [14,15], (2) the dual-continuum method, including double and multi-porosity, dual-permeability, or the more general multiple interacting continuum method (e.g. [1,5,11,17,19], and (3) the effectivecontinuum method (e.g. [18].

Among the most common models, Warren and Root [17] model is known as the pseudo-steady state model, since the matrix–fracture transfer term is expressed using a tanktype formulation. Streltsova's model (1983) is known as the pressure gradient model, since, according to her assumption, the matrix–fracture flow rate is controlled by the average pressure gradient in the matrix. The unsteady state models are developed by Kazemi [5], deSwaan [4], and Najurieta [9] models, where matrix-to-fracture flow is described by an unsteady state flow equation.

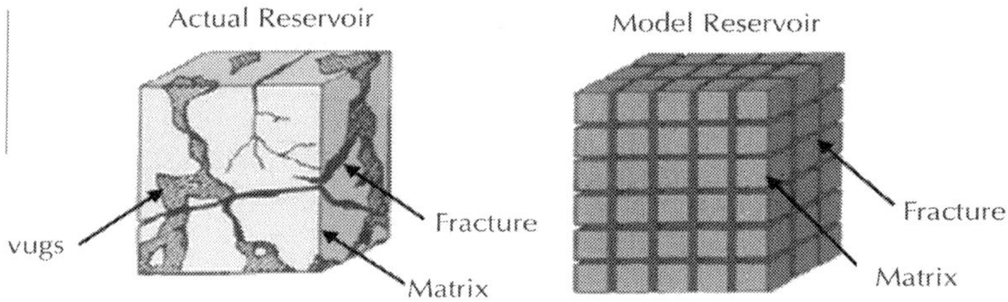

Figure 1: Warren and Root's idealization of NFRs (after [17]).

Table 1: Important features of some mathematical models describing flow from matrix to fracture

Warren and Root [17]
Analytical model
Pseudo-steady state model
Matrix flux is independent of spatial position and is proportional to the pressure difference between matrix and fracture
Simplifying the mathematical analysis of the flow problem
S-shaped transitional curve with an inflection point
The separation of the two parallel lines allows calculation of the storativity ratio
Kazemi [5]
Numerical model
Unsteady state model
Linear transitional curve with no inflection point
de Swaan [4]
Analytical model
Unsteady state model
A convolution theorem gives the relationship between the source term and the pressure
in the fracture medium
Linear transitional curve with no inflection point

Najurieta [9]
Analytical model
Unsteady state model
Approximate solution to de Swaan model
Only applicable to transient period (no boundary dominated period)
Linear transitional curve with no inflection point
Streltsova [16]
Analytical model
Pressure gradient model
Matrix flux is proportional to the averaged pressure gradient throughout the matrix block
Linear transitional curve with no inflection point

Considering the Warren and Root [17] model, the general assumptions in the well test analysis of NFRs are: (i) pseudo-steady state matrix flow the assumption that has been eliminated in this work by development of a numerical simulation model; (ii) the matrix consists of a set of porous rock systems that are not connected to each other, have a low transmissibility, and high storage capacity; (iii) the fracture system has low storage capacity, high transmissibility, and it interconnects the porous media; and (iv) the matrix supplies the fluids to the fractures, and the fractures transport the fluids to the well (i.e., the matrix does not provide fluid directly to the well).

MODELING

From the study of different models proposed for describing the flow from matrix to fracture it is concluded that although the Warren and Root model has good capabilities to predict the behavior of a NFR, it has several disadvantages:

- It is a pseudo- steady state model, the assumption that was used by the investigators to solve the obtained partially differential equation.

- Unlike other models proposed for NFRs as described in Table 1, the Warren and Root model has an S-shaped transitional curve with an inflection point, while existence of an S-shaped transition curve with an inflection point has not been reported by other investigators. This problem is mainly the result of replacement of the Bessel function obtained in the Laplace transform of the original PDE by a logarithmic approximation, since inversion of the Laplace transform of the original PDE could not be obtained by analytical methods.

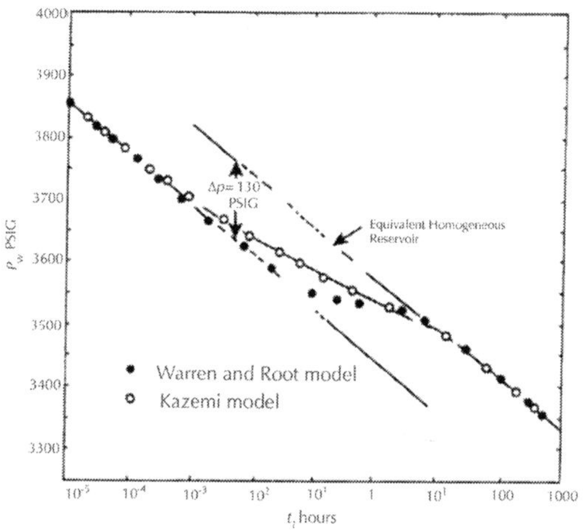

Figure 2. Pressure drawdown according to the model by Warren and Root (after [5]).

- As shown in Fig. 2, the Warren and Root model under predicts the length of transition section (between two straight lines), which is the result of assumptions used in this model.

In this work, authors used a numerical approach to modify the Warren and Root model. As it could be seen, the developed numerical method improves the results obtained by an analytical solution to the Warren and Root model especially in region 2. Developed computer program can simulate any test like, for example, conventional drawdown and buildup tests, injection test, interference test, and falloff test for a conventional reservoir or a NFR.

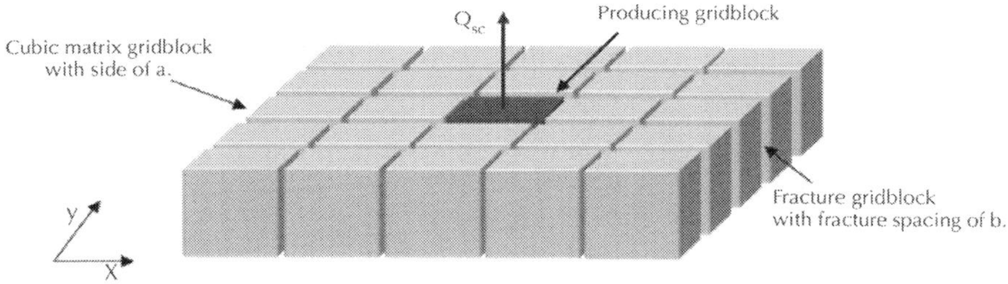

Figure 3: A typical 2D model of a NFR with 25 matrix gridblocks surrounded by fracture gridblocks.

Methodology

Fig. 3 shows a typical 2D model of a NFR which is used in this work, with 25 matrix gridblocks surrounded by fracture gridblocks.

In this work, a model similar to the model of conventional reservoirs is used. But a series of modifications have been made to convert the model of conventional reservoirs to the model of NFRs: (i) two different types of gridblocks called matrix and fracture gridblocks are used as shown in Fig. 3. (ii) These two types of gridblocks differ in porosity and permeability which is higher for fracture gridblocks compared to matrix gridblocks. (iii) Cross flow between gridblocks is defined using a transmissibility term. As it can be seen in Fig. 3, the size of network is controlled by the parameters, (i) a, side of cubic matrix gridblock, (ii) b, fracture spacing, (iii) n, number of matrix gridblocks in x or y direction.

Implicit Finite difference Formulation

Eq. (1) represents the diffusivity equation for a conventional reservoir in a two-dimensional (2D) directional model,

$$\frac{\partial}{\partial x}\left(\beta_c \frac{A_x k_x}{\mu_o B_o} \frac{\partial P}{\partial x}\right)\Delta x + \frac{\partial}{\partial y}\left(\beta_c \frac{A_y k_y}{\mu_o B_o} \frac{\partial P}{\partial y}\right)\Delta y + Q_{sc} = \frac{V_b}{\alpha_c} \frac{\partial}{\partial t}\left(\frac{\varnothing}{B_o}\right) \quad (1)$$

Values of b_c and a_c in the above equation, which are transmissibility conversion factor and volume conversion factor, respectively are given in Table 2.

Implicit finite difference formulation of Eq. (1) is as follows,

$$T_{x_{i+\frac{1}{2},j}}\left(P_{i+1,j}^{n+1}-P_{i,j}^{n+1}\right)-T_{x_{i-\frac{1}{2},j}}\left(P_{i,j}^{n+1}-P_{i-1,j}^{n+1}\right)+T_{y_{i,j+\frac{1}{2}}}\left(P_{i,j+1}^{n+1}-P_{i,j}^{n+1}\right)$$

$$-T_{y_{i,j-\frac{1}{2}}}\left(P_{i,j}^{n+1}-P_{i,j-1}^{n+1}\right)+Q_{sc_{i,j}}=\left(\frac{V_b ØC_t}{\alpha_c B\Delta t}\right)_{i,j}\left(P_{i,j}^{n+1}-P_{i,j}^{n}\right) \tag{2}$$

where transmissibility (T) is defined as,

Table 2: Values of b_c and a_c parameters

Parameter	Field unit	Metric unit
b_c	1.127	86.4×10^{-4}
a_c	5.614583	1

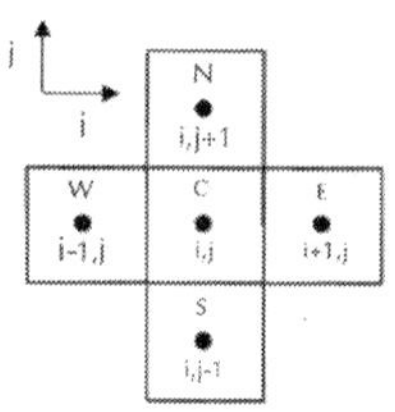

Figure 4: Naming convention for the matrix coefficients S, W, C, E and N.

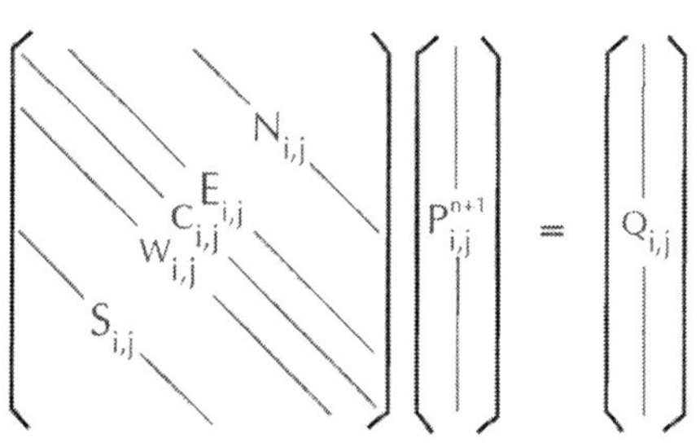

Figure 5: Schematic representation of the matrix equation generated by the 2D, slightly-compressible flow equation for a NFR.

$$T_{x_{i\pm\frac{1}{2},j}} = \left(\beta_c \frac{A_x k_x}{\mu B \Delta x}\right)_{i\pm\frac{1}{2},j} \qquad (3)$$

$$T_{y_{i,j\pm\frac{1}{2}}} = \left(\beta_c \frac{A_y k_y}{\mu B \Delta y}\right)_{i,j\pm\frac{1}{2}} \qquad (4)$$

In the model used in this work, there are two types of flow which are, (i) flow between a fracture gridblock and a matrix gridblock, (ii) flow between two fracture gridblocks. In the case of flow between a fracture gridblock and a matrix gridblock, the values of cross sectional area (A), permeability (k) and Dx are different, so average values are used in calculation. For Eqs. (3) and (4) the average values are calculated as follows,

Table 3: Data used in different runs of developed computer program for simulation of drawdown test in a NFR

Parameter	Base Case	Sensitivity Study
$n \times n$	51 × 51	11 × 11–71 × 71
a (ft)	100	10–300
b (ft)	1	0.1–10
φ_m	0.1	0.01–0.2
φ_m	1	0.1–1
k_m (md)	10^{-3}	10^{-3}–100
k_f (md)	500	100–10^5
B_o (bbl/stb)	1.2	1–2
μ_o (cp)	1	0.1–10
Q_{sc} (stb/day)	10	1–100
P_i (psi)	3000	–
c_o (psi^{-1})	10^{-5}	10^{-5}–10^{-2}
Δt (day)	0.1	0.001–1
t (day)	30	–

$$T_{fmx_{i\pm\frac{1}{2},j}} = \left(\frac{\beta_c}{\mu B}\frac{2a^2 k_m k_f}{ak_m + bk_f}\right)_{i\pm\frac{1}{2},j} \tag{5}$$

$$T_{fmy_{i,j\pm\frac{1}{2}}} = \left(\frac{\beta_c}{\mu B}\frac{2a^2 k_m k_f}{ak_m + bk_f}\right)_{i,j\pm\frac{1}{2}} \tag{6}$$

in which T_{fm} is the transmissibility between fracture and matrix gridblocks in x or y direction.

Two similar equations to (5) and (6) i.e. Eqs. (7) and (8) could be obtained for the case of transmissibility between two fracture gridblocks,

$$T_{ffx_{i\pm\frac{1}{2},j}} = \left(\frac{\beta_c}{\mu B}\frac{2abk_f}{a + b}\right)_{i\pm\frac{1}{2},j} \tag{7}$$

$$T_{ffy_{i,j\pm\frac{1}{2}}} = \left(\frac{\beta_c}{\mu B}\frac{2abk_f}{a + b}\right)_{i,j\pm\frac{1}{2}} \tag{8}$$

in which T_{ff} is the transmissibility between two fracture gridblocks.

Numerical Solution

For numerical solution of this model and evaluation of pressure distribution, Eq. (2) must be solved. This equation could be converted to the following form,

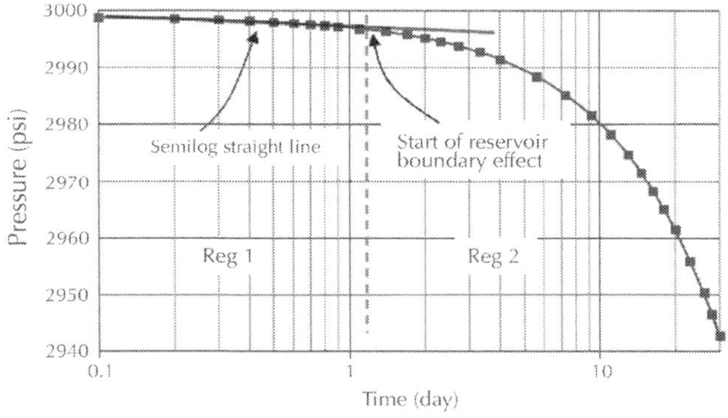

Figure 6: Pressure response versus time in a semilog plot for a conventional reservoir.

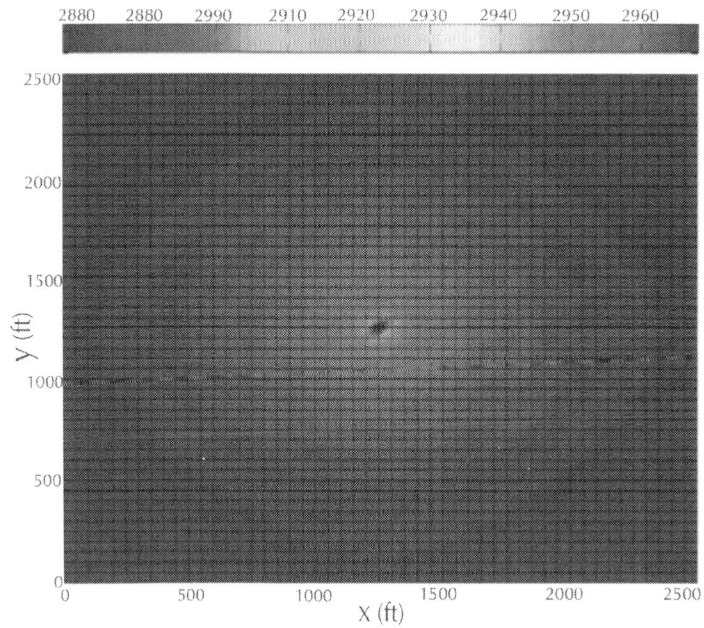

Figure 7: Pressure distribution in x–y plane in a typical 2D model of a conventional reservoir. As it can be seen in a conventional reservoir pressure distribution is smooth.

$$S_{i,j}P_{i,j-1}^{n+1} + W_{i,j}P_{i-1,j}^{n+1} + C_{i,j}P_{i,j}^{n+1} + E_{i,j}P_{i+1,j}^{n+1} + N_{i,j}P_{i,j+1}^{n+1} = Q_{sc_{i,j}} \quad (9)$$

in which,

$$S_{i,j} = T_{y_{i,j-\frac{1}{2}}} \quad (10)$$

$$W_{i,j} = T_{x_{i-\frac{1}{2},j}} \quad (11)$$

$$C_{i,j} = -\left[T_{y_{i,j-\frac{1}{2}}} + T_{x_{i-\frac{1}{2},j}} + T_{x_{i+\frac{1}{2},j}} + T_{y_{i,j+\frac{1}{2}}} + \left(\frac{V_b Øc_t}{\alpha_c B\Delta t}\right)_{i,j}\right] \quad (12)$$

$$E_{i,j} = T_{x_{i+\frac{1}{2},j}} \quad (13)$$

$$N_{i,j} = T_{y_{i,j+\frac{1}{2}}} \quad (14)$$

$$Q_{i,j} = -\left(\frac{V_b Øc_t}{\alpha_c B\Delta t}\right)_{i,j} P_{i,j}^{n} - Q_{sc_{i,j}} \quad (15)$$

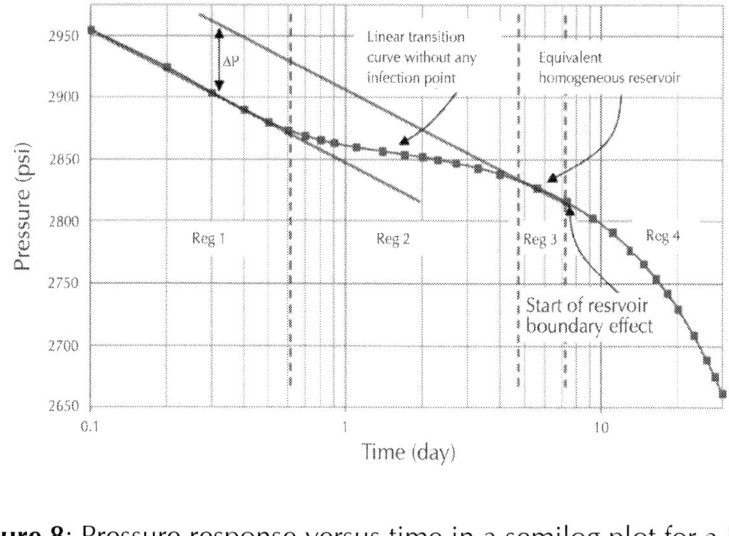

Figure 8: Pressure response versus time in a semilog plot for a NFR. Four different regions can be distinguished. Region 2 is a linear transition curve without any inflection point.

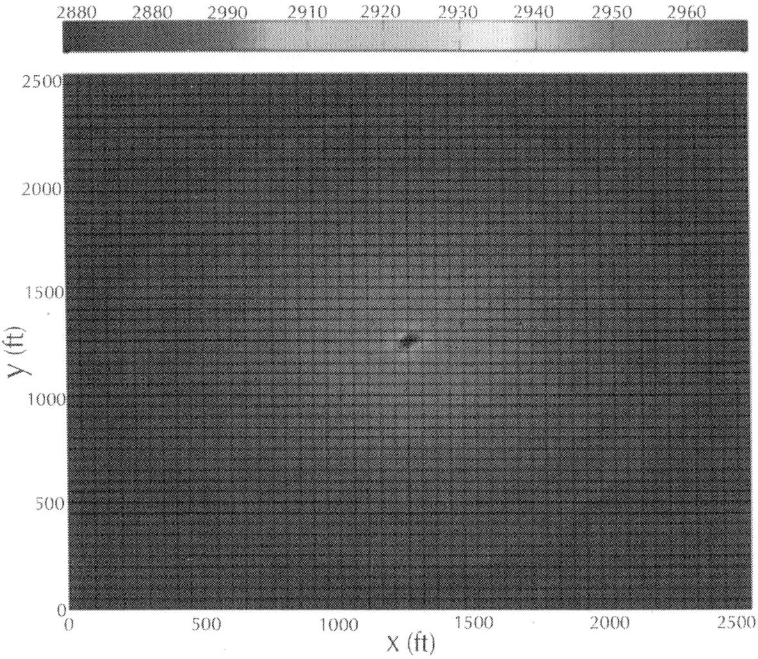

Figure 9: Pressure distribution in x–y plane in a typical 2D model of a NFR. In spite of a conventional reservoir, in a NFR pressure distribution is rough around the well and is smooth away from the wellbore.

We can now discuss the naming convention used for the coefficients in Eq. (9) and is shown in Fig. 4 which shows a typical gridblock in a 2D grid. For the orientation shown in this figure, the coefficients S, W, E, and N refer to the directions south, west, east, and north, respectively; the coefficient, C, refers to the grid cell itself (center).

Fig. 5 shows the matrix equation that results when Eq. (2) is written for each gridblock in the system.

SIMULATION RESULTS AND DISCUSSION

A computer program has been developed to solve the described model. The main data used in different runs of developed computer program are presented in Table 3.

By setting fracture parameters equal to the matrix parameters, the model of NFR is converted to the model of a conventional single porosity and single permeability reservoir. Figure 6 and Figure 7 show pressure response versus time in a plot and pressure distribution in the *x–y* plane in a typical 2D model of a conventional reservoir. As it can be seen in Fig. 6, two different regions in this plot can be determined. Region 1 in which, flow is through matrix gridblocks in a transient flow regime shows a straight line in semilog plot. In region 2 the pressure disturbance has been reached to the boundary and it deviates from the semilog straight line. Also it should be noted that for a real conventional reservoir, before region 1 there is another deviation from the straight line due to the wellbore storage effect, but since this model is in 2D this region does not appear.

In Fig. 7, pressure distribution in the *x–y* plane of a 2D conventional reservoir is shown, where the producing block is considered in the center of model. As it could be seen in this figure, pressure distribution in a conventional reservoir is smooth.

There are two regions called regions 1 and 2. In region 1 the flow regime is transient and in region 2 the flow regime is a pseudo-steady state.

Following is the discussion on simulation results for a typical NFR. It can be seen in Fig. 8, in a semilog plot of pressure versus time for a typical NFR, four different regions can be distinguished: (i) region 1, which is the first straight line and it represents the homogeneous behavior of the naturally fractured medium before the matrix medium starts to respond (transient radial flow). The slope of this line gives the fracture permeability; (ii) a transition section (between two straight lines), which corresponds to the onset of interporosity flow. It should be noted that in this work the transition curve is linear without any inflection point as predicted by some investigators [5], [4], [9] and [16]; (iii) the second semilog straight line, which represents the homogeneous behavior of composite media (fracture permeability with the sum of matrix and fracture storages) when recharge from the matrix medium is fully established; (iv) the fourth region which may exist in the semilog plot of a typical NFR showing a deflection from the second semilog straight line which is called the reservoir boundary effect.

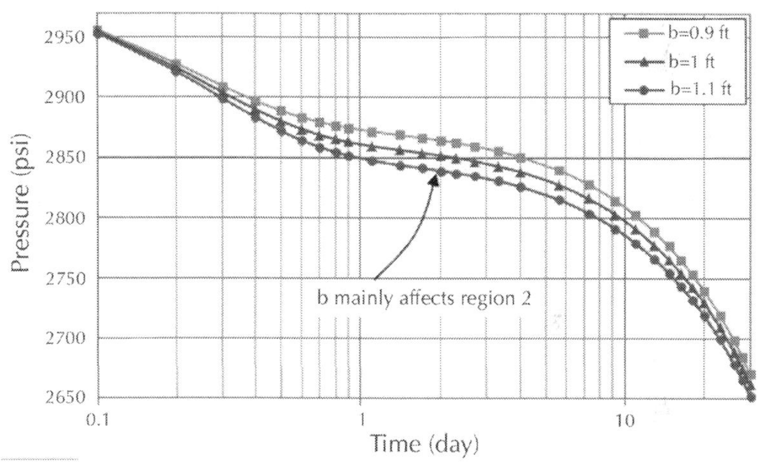

Figure 10: Effect of fracture spacing on semilog plot of a NFR. Fracture spacing is more important for region 2 which results in more pressure drop and an increase in the length of transition zone.

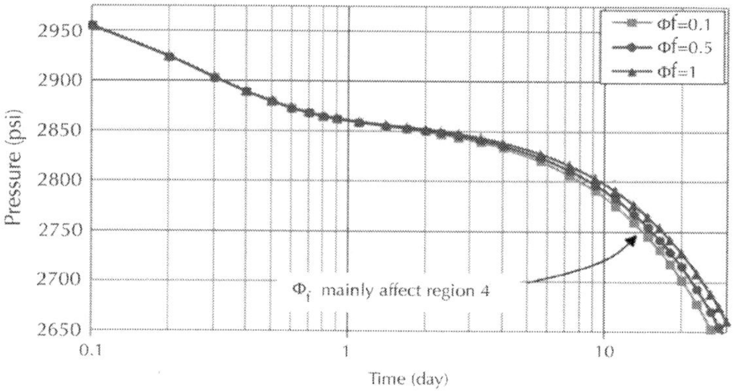

Figure 11: Effect of fracture porosity on semilog plot of a NFR. Fracture porosity is more important for region 4. In this region a large volume of fractures contributes to flow, which causes less pressure drop.

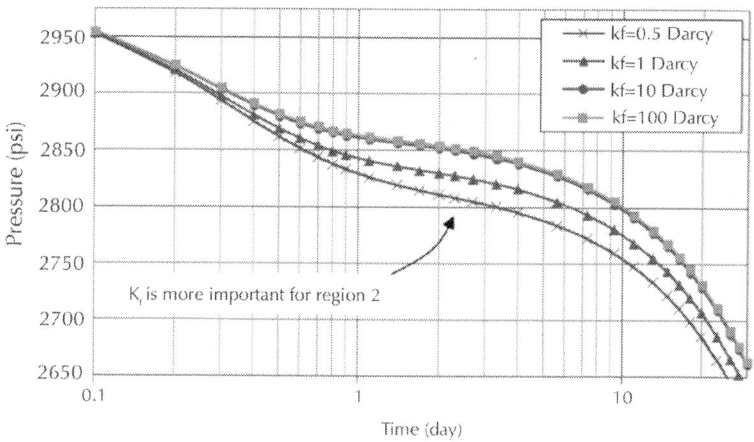

Figure 12: Effect of fracture permeability on semilog plot of a NFR. Fracture permeability is more important for region 2 due to the onset of interporosity flow which results in a less pressure drop and a decrease in the length of transition zone.

As it can be seen in Fig. 9 in a NFR, pressure distribution is rough around the well and is smooth away from the wellbore. This behavior is the result of a great contrast in the matrix and fracture gridblock's porosity and permeability which causes to flow be mainly from the fracture gridblocks and consequently matrix gridblocks have a less contribution to flow especially in regions 1 and 2.

A parametric study is done in this work to investigate the effect of matrix and fracture gridblocks on the semilog plot of a typical NFR. Table 3 represents the main parameters which are subjected to this study. Among these parameters matrix and fracture parameters which are more important will be discussed.

Fig. 10 shows the effect of fracture spacing on semilog plot.

As it can be seen, the size of fracture spacing is more important for region 2 which is a transition zone. It is due to the onset of interporosity flow that occurs in this region. More fracture spacing results in a more pressure drop and an increase in the length of transition zone.

Figure 11 and Figure 12 represent the effect of fracture porosity and permeability on the semilog plot of a typical NFR. Fracture porosity is more important for region 4 since the volume of fractures is less compared to the volume of matrix gridblocks and in this region where a large volume of fractures contribute to flow, it causes a less pressure drop.

According to Fig. 12, fracture permeability in all regions except region 1 is important and it is more important for region 2 due to onset of interporosity flow. An increase in k_f results in less pressure drop and a decrease in the length of transition zone. Also at high values of k_f, the effect of increase in k_f on the behavior of a NFR decreases.

After the investigation of fracture parameters, the focus is on the effect of matrix parameters on the behavior of a typical NFR. Fig. 13 represents the effect of matrix porosity on the behavior of a typical NFR. As it can be seen, matrix porosity affects the behavior of a NFR in all regions except region 2 in which there is onset of interporosity flow. Also for region 4 its effect is more due to the contribution of a large volume of matrix gridblocks in flow.

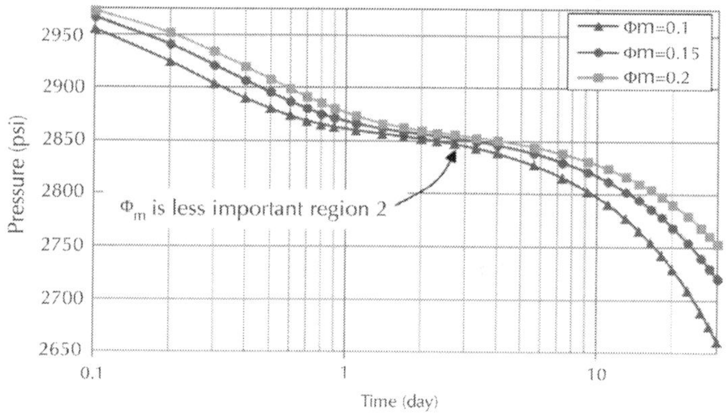

Figure 13: Effect of matrix porosity on semilog plot of a NFR. Matrix porosity is more important for region 4 due to the contribution of a large volume of matrix gridblocks in flow. It is less important for region 2 in which there is an onset of interporosity flow.

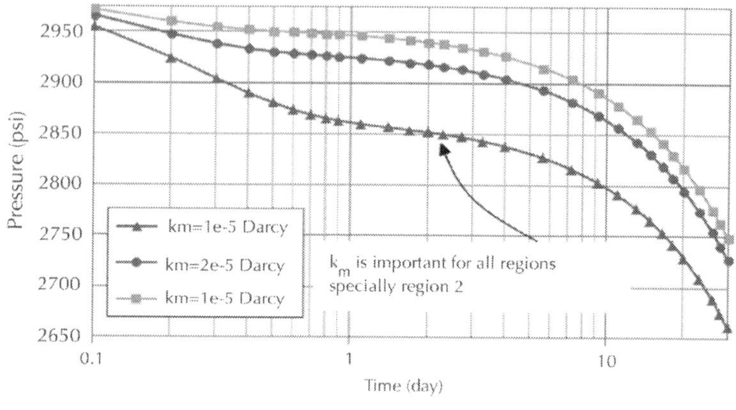

Figure 14: Effect of matrix permeability on semilog plot of a NFR. Matrix permeability is important for all regions except region 1 in which flow is only through fractures.

The effect of matrix permeability on the behavior of a typical NFR is illustrated in Fig. 14. Matrix permeability is important for all regions and is less important for region 1 in which flow is only through fractures. In region 2 its effect is more due to the effect of interporosity flow. Also according to Figure 13 and Figure 14, in a typical NFR more matrix porosity and permeability causes less pressure drop and a decrease in the length of transition zone.

CONCLUSIONS

A mathematical model similar to the model of conventional reservoirs was presented for simulation of a slightly compressible flow in a typical NFR based on Warren and Root model. Well testing was drawdown whereas the computer program is capable to simulate any other tests. To convert the model of conventional reservoirs to the model of NFRs, two types of gridblocks called matrix and fracture gridblocks were defined which differ in porosity and permeability. Then two types of transmissibility terms were defined between matrix and fracture gridblocks to model a NFR. Two drawdown simulations were done for a conventional reservoir and a NFR. In the case of NFR, in spite of the Warren and Root analytical solution, results of this work show a linear transition curve without any inflection point as reported by other investigators. In the semilog plot of a NFR, at most four regions can be distinguished. A comprehensive parametric study on matrix and fracture was done to investigate the influence of these parameters in four regions. Region 1 is mainly affected by matrix parameters, *i.e.*, matrix porosity and permeability and can be influenced by fracture spacing and permeability. Behavior of this region is independent of fracture porosity. Except the matrix and fracture porosity, other parameters influence the pressure drop in region 2 and change its length. Region 3 is influenced by all the matrix and fracture parameters especially by permeability values. In region 4, matrix parameters are more important and fracture spacing influence is negligible.

REFERENCES

1. G.I. Barenblatt, Yu.P. Zheltov, Soviet Phys., Doklady 5 (1960) 522.
2. B. Berkowitz, Adv. Water Resour. 25 (2002) 861–884.
3. C.C. Chen, K. Serra, A.C. Reynolds, R. Raghavan, Soc. Petrol. Eng. J. 25 (1985) 451–464.
4. A. deSwaan, Soc. Petrol. Eng. J. (1976) 117–122, Trans. AIME, p. 228.
5. H. Kazemi, Soc. Petrol. Eng. J. (1969) 451–462, Trans. AIME, p. 261.
6. C.H. Lai, G.S. Bodvarsson, C.F. Tsang, P.A. Witherspoon, in: Paper SPE 11688 Presented at the 1983 SPE California Regional Meeting, Ventura, 1983.
7. M.J. Mavor, H. Cinco, in: Paper SPE 7977 Presented at the 1979 SPE California Regional Meeting, Ventura, 1979.
8. G.H. Montazeri, S.A. Tahami, B. Moradi, E. Safari, Brazilian Journal of Petroleum and Gas 5(3) (2011) 139–147.
9. H.L. Najurieta, J. Petrol. Technol. (1980) 124–150.
10. A.S. Odeh, Soc. Petrol. Eng. J. Trans. AIME 234 (1965) 60–66.
11. K. Pruess, T.N. Narasimhan, Soc. Petrol. Eng. J. 25 (1985) 14–26.
12. M.A. Sabet, Well Test Analysis, Gulf Publishing Company, 1991.
13. K. Serra, A.C. Reynolds, J. Petrol. Technol. (1983) 2271–2283.
14. D.T. Snow, Ph.D. Dissertation, University of California, Berkeley, 1965, p. 331.
15. S. Stothoff, D. Or, in: B. Faybishenko, P.A. Witherspoon, S.M. Benson (Eds.), Dynamics of fluids in fractured rocks, concepts and recent advances, AGU Geophysical Monograph 122, Washington, DC, 2000, 269–279.
16. T.D. Streltsova, Soc. Petrol. Eng. J. (1983) 769–780.
17. J.E. Warren, P.J. Root, Soc. Petrol. Eng. J. (1963) 245–255,

Trans. AIME, p. 228.

18. Y.S. Wu, in: B. Faybishenko, P.A. Witherspoon, S.M. Benson (Eds.), Dynamics of Fluids in Fractured Rocks, Concepts and Recent Advances, AGU Geophysical Monograph 122, Washington, DC, 2000, pp. 299–312.
19. Y.S. Wu, K. Pruess, SPE Reservoir Eng. 3 (1988) 327–336.

Chapter 3

A Sequential Implicit Algorithm of Chemo-thermo-poro-mechanics for Fractured Geothermal Reservoirs

Jihoon Kim[a, b], Eric Sonnenthal[a],
Jand onny Rutqvist[a]

[a]Earth Sciences Division, Lawrence Berkeley National Laboratory, 1 Cyclotron Road 74R316C, Berkeley, CA 94720, USA

[b]Harold Vance Department of Petroleum Engineering, Texas A&M University, 3116 TAMU 407K Richardson Building College Station, TX 77843, USA

ABSTRACT

We describe the development of a sequential implicit formulation and algorithm for coupling fluid-heat flow, reactive transport, and geomechanics. We consider changes in pore volume from dissolution caused by chemical reactions, in addition to coupled flow and geomechanics. Moreover, we use the constitutive equations for the multiple porosity model for fractured geothermal reservoirs, employing failure-dependent permeability dynamically and updating it every time step. The proposed sequential algorithm is an extension of the fixed-stress split method to chemo-thermo-poro-mechanics, facilitating the use of existing flow-reactive transport and geomechanics simulators.

We first validate a simulator that employs the proposed sequential algorithm, matching the numerical solutions with the analytical solutions such as Terzaghi's and Mandel's problems for poro-mechanics and the reference solutions of chemo-poro-mechanics and chemo-thermo-poro-mechanics in the 1D elastic problems. We also perform convergence test, and the proposed algorithm shows fast convergence, when full iteration is taken, and first order accuracy in time for the staggered approach.

We then investigate two test cases: 2D multiple porosity elastic and 3D single porosity elastoplastic problems, and explore the differences in coupled flow and geomechanics with and without reactive transport. We find that the change in pore-volume induced by mineral dissolution can impact on fluid pressure and failure status, followed by significant changes in permeability and flow variables, showing strong interrelations between flow-reactive transport and geomechanics.

INTRODUCTION

Enhanced geothermal reservoirs that exhibit increased permeability resulting from shear stimulation of a fracture volume can provide large heat extraction accompanied by economic flow rates and energy production. Fractures are highly compressible compared with intact rock, so they may affect overall geomechanical responses significantly, even though they occupy a smaller bulk volume (Bai, 1999). Permeability can significantly be affected by deformation of fractures through the relationship between the fracture aperture and permeability (Snow, 1965 and Rutqvist and Stephansson, 2003). Furthermore, chemical or thermal nonequilibrium between injection fluid and the reservoir host rock can result in dissolution and precipitation, changing the porosity (Xu et al., 2011). Such changes in pore volume can change fluid pressure, and hence effective stress and strain. Permeability changes through mineral-water reactions and geomechanics affect solute transport which can lead to strong thermal, hydrological, mechanical, and chemical coupling. These physical processes occur simultaneously, but they have different time scales. The geomechanical problem is written as the elliptic partial differential equation (PDE) for modeling quasi-static mechanics, showing the instantaneous mechanical responses (Coussy, 1995 and Lewis and Schrefler, 1998), while pressure and saturation equations of the multiphase flow problems are expressed as the parabolic and hyperbolic PDEs, which model time-dependant physical behavior (Aziz and Settari, 1979 and Pruess et al., 1999). Reaction between fluid and the porous media can be modeled based on either a rate-independent equation from local equilibrium assumption or a rate-dependent form of kinetic reaction (Moridis et al., 2008 and Xu et al., 2008). The different time scales and strong couplings in those physics result in high numeral stiffness and strong nonlinearity, when we solve the coupled mathematical equations numerically, thus robust and rigorous numerical modeling is necessary for accuracy.

The geothermal reservoirs are naturally or artificially fractured for economic feasibility, consisting of the fracture(s) and rock matrix. The multiple continuum approach is a widely used approach that can represent the fracture-rock matrix systems consisting of a highly permeable fracture continuum transporting fluid over the domain, while the matrix stores fluid and conveys it to the highly permeable fracture continuum (Barenblatt et al., 1960, Berryman, 2002, Sonnenthal et al., 2005 and Taron et al., 2009). In geothermal reservoirs, we may introduce more than two continua (e.g., representing fracture and several rock matrix continua) for more accurate modeling of heat flow. Recently, Kim et al. (2012) proposed formulation and sequential numerical algorithms for coupled fluid-heat flow and geomechanics within such a multiple continuum approach. We have implemented these functionalities within ROCMECH, an in-house geomechanics simulator, which is coupled to TOUGH family codes, flow simulators.

The study presented in this paper is based on the latter study, employing nonlinear dynamic strain and failure-dependent permeability in order to capture complex interactions between flow and geomechanics. In this study, we consider changes in pore volume, induced by chemical reaction, when solving for fluid and heat flow. When combining the two simulators, we solve fluid and heat flow, geomechanics, and reactive transport problems sequentially, adopting implicit solution schemes for the subproblems (i.e., sequential implicit method). Specifically, when solving flow and geomechanics, we employ the fixed-stress sequential method, which solves flow first, fixing the total stress field locally, and solves geomechanics at the next step from the calculated flow solutions. The fixed-stress method can provide numerical unconditional stability and high accuracy, comparable to the fully implicit method, regardless of the coupling strength in pore volume between flow and geomechanics (Kim et al., 2011b, Kim et al., 2013 and Kim et al., 2012). Taron et al. (2009) employed the undrained sequential method, which solves geomechanics first, fixing fluid mass locally, and then solves flow at the next step. The undrained sequential method can also provide numerical stability in solving flow and geomechanics (Armero, 1999), but causes inaccuracy or non-convergence for the high coupling strength in the case of highly deformable porous media (e.g., fracture) and incompressible fluid (e.g. liquid water) (Kim et al., 2011a). The fixed-stress method can easily be implemented by using the porosity function and its correction (Settari and Mourits, 1998 and Kim et al., 2012). In this study, we incorporate the porosity change from chemical reaction into the sequentially coupled simulator, TOUGHREACT+ROCMECH (shortly, T+M). In T+M, flow-geomechanics properties and variables (e.g., porosity, permeability, fluid-phase pressure and saturation, displacement, effective stress, failure status) are updated dynamically along with the transient coupled simulation. We perform several verification test cases for

T+M (e.g., the Terzaghi and Mandel problems, 1D elastic chemo-poro-mechanics and chemo-thermo-poro-mechanics problems), as well as their convergence analysis. Then, we investigate 2D and 3D geothermal problems, using coupled thermo-poro-mechanics simulations with and without reactive transport, and analyze differences between them.

MATHEMATICAL DESCRIPTION

We briefly describe the governing equations of geomechanics, fluid and heat flow with reactive solute transport. Next, we introduce constitutive equations for the multiple porosity model and the changes in pore volume from geomechanics and chemical reactions.

Governing Equations

The governing equation for geomechanics is based on the quasi-static assumption, written as (Coussy, 1995)

$$\mathrm{Div}\,\boldsymbol{\sigma} + \rho_b \mathbf{g} = \mathbf{0}, \tag{1}$$

where Div is the divergence operator, $\boldsymbol{\sigma}$ is the total stress tensor, ρ_b is the bulk density, and g is the gravity vector. The infinitesimal transformation is used to allow the strain tensor, $\boldsymbol{\varepsilon}$, to be the symmetric gradient of the displacement vector, **u**,

$$\boldsymbol{\varepsilon} = \frac{1}{2}(\mathbf{Grad}\,\boldsymbol{u} + \mathbf{Grad}^t\,\boldsymbol{u}), \tag{2}$$

where Grad is the gradient operator, and Grad^t u is the transpose of Grad u. From here on, tensile stress and strain are positive.

The governing equations for multiphase-multicomponent fluid and heat flow are derived from the conservation law (Pruess et al., 1999), as follows:

$$\frac{d}{dt}\int_{\Omega} m^k d\Omega + \int_{\Gamma} \mathbf{f}^k \cdot \mathbf{n}\, d\Gamma = \int_{\Omega} q^k d\Omega, \tag{3}$$

where the superscript k indicates the fluid component or heat. d(·)/dt means the time derivative of a physical quantity (·) relative to the motion of the solid skeleton. mk is mass of component k or heat, where the superscript $(\cdot)^\theta$ indicates heat. $\mathbf{f}^k$ and q^k are its flux and source terms on the domain Ω with a boundary surface Γ, respectively, where **n** is the normal vector to the boundary.

The mass flow term $\mathbf{f}^k$ in Eq. (3) is given by

$$\mathbf{f}^k = \sum_J (\mathbf{w}_J^k + \mathbf{J}_J^k), \tag{4}$$

where w^k_j and j^k_j are the convective and diffusive mass flows of component k in the phase J . When the fluid phase is liquid, w^k_j is described by Darcy's law as

$$\mathbf{w}_J^k = \psi_J^k\, \mathbf{w}_J$$
$$\mathbf{w}_J = -\frac{\rho_J\, k_{rJ}}{\mu_J}\, \mathbf{k}_p\, (\mathbf{Grad}\, P_J - \rho_J\, \mathbf{g}), \tag{5}$$

where $_{kp}$ is the second order absolute permeability tensor; $_{\rho J}$, $_{\mu J}$, $_{krJ}$, and PJ are the density, viscosity, relative permeability, and pressure of the fluid phase J, respectively. ψJk is the mass fraction of component k in the phase J . In the case that the fluid is gaseous, w^k_j can be written as

$$\mathbf{w}_J = -\left(1 + \frac{k_K}{P_J}\right) \mathbf{k}_p\, \frac{\rho_J\, k_{rJ}}{\mu_J}\, (\mathbf{Grad}\, P_J - \rho_J\, \mathbf{g}), \tag{6}$$

where kK is the Klinkenberg factor (Klinkenberg, 1941). The diffusive flow j^k_j is described as

$$\mathbf{J}_k^J = -\,\Phi\, S_J\, \tau_G\, \mathbf{D}_J^k\, \rho_J\, \mathbf{Grad}\, \psi_J^k, \tag{7}$$

where D^k_j is the second order hydrodynamic dispersion tensor of component k in the phase J , and τ_G is the gas tortuosity. Φ (Lagrange›s porosity, also called reservoir porosity) is defined as the ratio of the pore volume in the deformed configuration to the bulk volume in the reference (typically initial) configuration. SJ is the saturation of the fluid phase J.

The heat flux, $\mathbf{f}^\theta$, is obtained by heat conduction and convection laws, written as

$$f^\theta = -\left((1-\Phi)\mathbf{K}_R + \Phi S_J \mathbf{K}_J\right)\mathbf{Grad}\, T + h_J \mathbf{w}_J, \tag{8}$$

where K_R and K_J are the second order thermal conductivity tensors of the porous media and fluid phaseJ, respectively. T is temperature. hJ is the specific enthalpy of phase J.

The governing equation for solute transport with chemical reaction (i.e., reactive transport) is obtained from mass conservation (Xu et al., 2011), described as

$$\frac{d}{dt}(\Phi C_i) + \mathrm{Div}(\Phi \mathbf{v}_i C_i - \Phi \mathbf{D}_i \mathbf{Grad}\, C_i) = -\sum_{m=1}^{n_c} \gamma_{im} r_m + q_i, \tag{9}$$

where Ci is the concentration of the i th chemical. v_i and D_i are the Darcy velocity vector and diffusion coefficient matrix of the ith chemical, respectively. γ_{im} and rm are the stoichiometric coefficient of the ith chemical in the mth chemical and the reaction rate of the mth chemical, respectively. nc is the total number of the chemicals, and q_i is the possible external source of the ith chemical.

Constitutive Relations in Coupled Flow and Geomechanics

The constitutive relations for thermoporomechanics in a single porosity model are originally based on the Biot theory (e.g., Biot, 1941, Biot and Willis, 1957, Geertsma, 1957 and Coussy, 1995). In this study, we employ the constitutive equations proposed by Kim et al. (2012) for the multiple porosity model (Fig. 1), developed from the isothermal double porosity elastic model (Berryman, 2002) into the nonisothermal multiple porosity elastoplastic model, described as

$$\delta\sigma = \overbrace{\mathbf{C}_{up}: \delta\underbrace{(\boldsymbol{\varepsilon} - \boldsymbol{\varepsilon}_p)}_{\varepsilon_e}}^{\delta\sigma'} - b^*_{lJ}\delta p_{l,J}\mathbf{1} - K_{dr}\bar{b}_l\delta T\mathbf{1},\ b^*_{lJ} = -K_{dr}(bS_J)_l, \tag{10}$$

$$\frac{1}{K_{dr}} = \frac{\eta_k}{K_k},\ \mathbf{C}_{up} = K_{dr}\left(\frac{\eta}{K}\right)_k \mathbf{C}_k,\ b_l = -\left(\frac{\alpha\eta}{K}\right)_l,\ \bar{b}_l = 3(\alpha_T\eta)_l, \tag{11}$$

$$\delta\zeta_{(l,J)} = b^*_{lJ}\delta\varepsilon_v + L^{-1}_{lJ,m,I}\delta p_{m,I} - D_{lJ,m}\delta T_m, \tag{12}$$

$$\left(\bar{S} - \bar{s}_J\delta m_J\right)_l = -\bar{b}_l K_{dr}\delta\varepsilon_v - \bar{D}_{l,m,I}\delta p_{m,I} + \bar{D}_{l,m}\delta T_m, \tag{13}$$

where double indices indicate summation. The subscript l indicates a material (sub-element) within a gridblock. δ implies variation relative to the motion of the solid skeleton. $\boldsymbol{\sigma}'$, Kdr, and $_{\text{Cup}}$ are the effective stress tensor, drained bulk modulus, and elasticity tensor at the level of the gridblock, respectively. **1** is the rank-2 identity tensor. $\delta_{\zeta(l,J)}=\delta_{m(l,J)}/_{\rho(l,J)}$, where $\delta_{m(l,J)}$ and $_{\rho(l,J)}$ are mass and density of fluid phase J in the material l, respectively. εv is the total volumetric strain. $b_{lJ,}{}^*$ and $\bar{b}_1$ are the coupling coefficients. αl, $\alpha_{(}$T,l$_{)}$, ηl, and Kl are the Biot coefficient, thermal dilation coefficient, volume fraction, and drained bulk modulus for material l, respectively. α_1=1-K_l / K_l^s, where Ksl is the intrinsic solid grain bulk modulus of material l . $_{\text{Cl}}$ is the isothermal drained elasticity tensor for material l . $_{\text{Ll,J,m,I}}$represents a modulus for the multiple porous model, similar to the Biot modulus matrix, and readers can refer to Kim et al. (2012) for the specific forms of $_{\text{Ll,J,m,I}}$. is the total entropy, and $\bar{S}_J$ is the internal entropy per unit mass of phase J (i.e., specific entropy of phase J). $\bar{D}_{l,m,I}$ is the coupling coefficient between fluid flow and heat flow based on thermodynamics. $\tilde{D}_{l,m}$ represents relations between temperatures of all materials in the heat flow simulation (i.e., total heat capacity matrix).

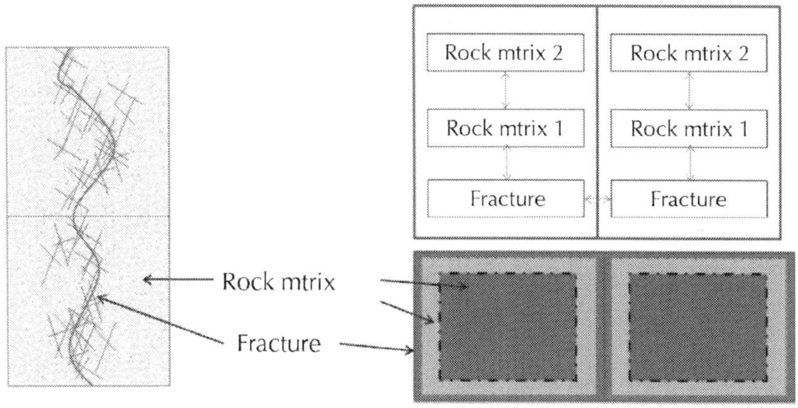

Figure 1: Left: A schematic diagram of a fracture-matrix system. Right: A conceptual model of the multiple interacting continuum (MINC) model (Pruess and Narasimhan, 1985). In the MINC model, fluid flows though the fracture medium over the domain. On the other hand, the rock matrix medium stores fluid and conveys it to the fracture medium. We may have several rock matrix media for accurate modeling of fluid-heat flow.

The local effective stress and local total strain at each subelement l (σ'_l and ${}_l$, respectively) are obtained as

$$\delta\sigma'_l = (\mathbf{C}:\delta\boldsymbol{\varepsilon})_l,\ \delta\varepsilon_l = \frac{K_{dr}}{K_k}\delta\boldsymbol{\varepsilon},\ \delta\boldsymbol{\sigma}' = \sum_{l=1}^{n_m}\eta_l\delta\sigma'_l, \tag{14}$$

where nm is the total number of the materials.

In this study, we neglect the heat contribution (i.e., entropy production) directly from geomechanics to heat flow, ignoring the term related to $-\tilde{b}_l K_{dr} \delta_{\varepsilon_v}$ of Eq. (13). This assumption is justified when heat capacity of material or fluid is high, or when direct heat generation from deformations is negligible (Lewis and Schrefler,

1998). Then, according to Kim et al. (2012), Lagrange's porosity of material l without chemical reaction, Ml, can be written as

$$\delta\Phi_l^{\ M} = \left(\frac{\alpha_l^2}{K_l} + \frac{\alpha_l - \Phi_l}{K_s}\right)\delta p_l + 3\alpha_{T,l}\alpha_l\delta T_l - \frac{b_l}{\eta_l}\delta\sigma_v, \tag{15}$$

where σv is the total (volumetric) mean stress. Eq. (15) is typically known as the porosity change, when reservoir simulation is coupled to geomechanics. The additional change in the porosity of material l due to reactive solute transport, $\delta\Phi^c{}_l$, can be formulated as (Xu et al., 2006)

$$\delta\Phi_l^{\ C} = -\sum_{m=1}^{n_c} \delta\chi_{m,l}, \tag{16}$$

where $\chi_{m,l}$ is the mth solid volume fraction of material l. Then, from Eqs. (15) and (16), we obtain the total porosity change due to geomechanics and geochemistry, as follows.

$$\begin{aligned}\delta\Phi_l &= \delta\Phi_l^{\ M} + \delta\Phi_l^{\ C} \\ &= \left(\frac{\alpha_l^2}{K_l} + \frac{\alpha_l - \Phi_l}{K_s}\right)\delta p_l + 3\alpha_{T,l}\alpha_l\delta T_l - \frac{b_l}{\eta_l}\delta\sigma_v - \sum_{m=1}^{n_c} \delta\chi_{m,l}.\end{aligned}$$

NUMERICAL INTEGRATION

We use finite volume and finite element methods for flow and geomechanics for space discretization, respectively. As indicated in Kim et al. (2012), this mixed finite volume and finite element formulation has local conservations of mass and heat at the element level for flow, continuous displacement field in geomechanics, and convergent approximations in space with the lowest order discretization. Furthermore, the mixed formulation shows relatively stable approximations because of a piecewise constant interpolation for fluid pressure, compared with the finite element method for flow that is formulated based on a piecewise continuous interpolation for pressure (Vermeer and Verruijt, 1981, Phillips and Wheeler, 2007, Kim et al., 2011a and Kim et al., 2011b). In time discretization, the backward Euler method is used as typically employed in reservoir simulation (Aziz and Settari, 1979). The initial total stress satisfies the geomechanical equilibrium, consistent with the boundary conditions.

In the numerical integration, we use a sequential implicit approach, by which fluid and heat flow, geomechanics, and reactive solute transport are solved sequentially, with each subproblem being solved implicitly, as shown in Fig. 2. TOUGHREACT, a coupled flow and reactive solute transport simulator, is also based on the sequential implicit approach (Xu et al., 2008). We develop the modified fixed-stress-split method in coupling between fluid-heat flow and geomechanics for the multiple porosity model, incorporating a chemo-thermo-poro-mechanics approach. Specifically, from Eq. (17), we can easily implement the modified fixed stress split by using the porosity function and its correction, as follows.

$$\Delta\Phi_l^n = \left(\frac{\alpha_l^2}{K_l} + \frac{\alpha_l - \Phi_l^n}{K_s}\right)\Delta p_l^n + (3\alpha_{T,l}\alpha_l)\Delta T_l^n - \frac{b_l}{\eta_l}\Delta\sigma_v^{n-1} + \Delta\Phi_l^{C,n-1}, \tag{17}$$

where $\Delta(\cdot)^n=(\cdot)^{n+1}-(\cdot)^n$. The fixed stress split is unconditionally stable (Von Neumann stability and B-stability for linear and nonlinear problems, respectively) and accurate in coupled flow and geomechanics without the reactive transport (Kim et al., 2011b, Kim et al., 2013, Kim et al., 2012 and Kim and Moridis, 2014). However, unconditional stability of chemo-thermo-poro-mechanics has not been proved mathematically, because thermo-poro-mechanics is only a subproblem of chemo-thermo-poro-mechanics.

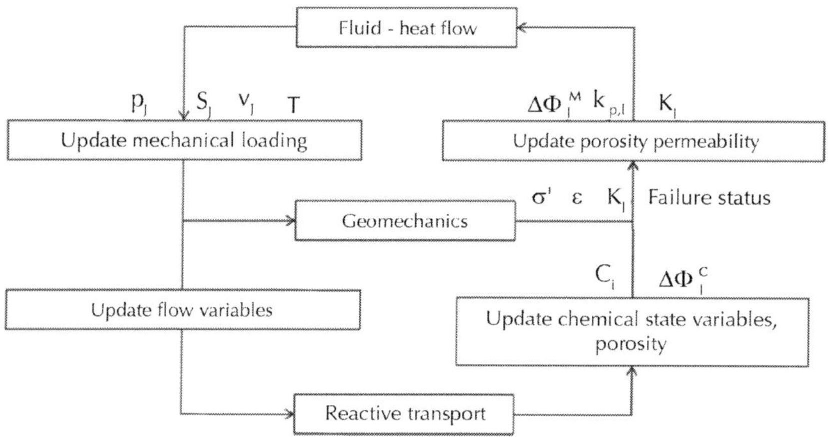

Figure 2: The sequential implicit algorithm of chemo-thermo-poro-mechanics. Flow, geomechanics, reactive transport simulators are communicated sequentially.

Once individual parallel codes for respective flow and geomechanics are available, the proposed algorithm can easily be applied to construct a parallelized coupled simulator, because it only requires simple and local code modification with data communication gridblock by gridblock. Thus, we can keep high scalability if the individual parallel codes are highly scalable.

NUMERICAL EXAMPLES

We first conduct tests of verification and numerical convergence for T+M's chemo-thermo-poro-mechanics approach. To this end, we first introduce the Terzaghi and Mandel problems to verify the simulator in poromechanics. Then we compare TOUGHREACT only (TR-only) with T+M for 1D elastic single-phase flow problems, because the 1D elastic problems with no extra external mechanical loading can allow the decoupling of flow and geomechanics, when we use an exact pore compressibility (i.e., (1/)(d /dp)) and thermal expansivity (i.e., (1/)(d /dT)) in flow. We also perform convergence analysis for these test cases, such as number of iterations for full iteration and accuracy in time. After verification and convergence analysis, we compare the numerical results among chemo-thermo-poro-mechanics, thermo-poro-mechanics without reactive transport, and flow-only with and without reactive transport for naturally and artificially fractured geothermal reservoirs.

Verification and Numerical Convergence

Terzaghi's and Mandel's Problems

Fig. 3 shows the domains for the (isothermal) Terzaghi and Mandel problems, in which 2D plane strain geomechanics is employed. For Terzaghi's problem, we have 31 gridblocks, the sizes of which are uniform, 1.0 m. We impose a drainage boundary on the left side and no-flow conditions on the right side. Liquid water is fully saturated, and the initial and boundary pressures are both 8.3 MPa. The initial total stress is also 8.3 MPa over the domain, and we apply a 16.6 MPa normal to the left boundary, two times greater than the initial total stress. The Young's modulus and Poisson ratio's are 450 MPa and 0.0, respectively. Only horizontal displacement is allowed. The monitoring well is located at the right end.

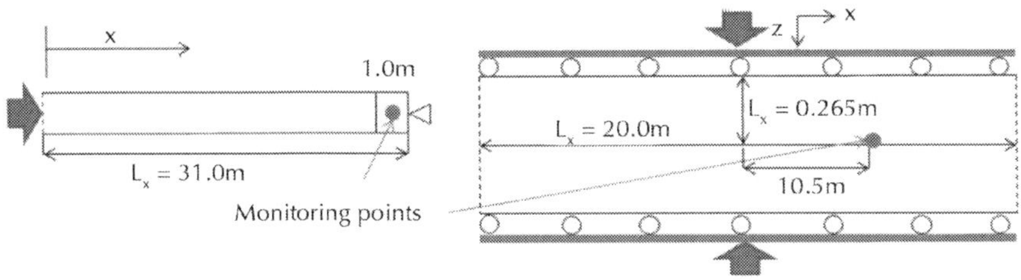

Figure 3: Left: Terzaghi's problem. Right: Mandel's problem. Verification for poromechanical effects is tested.

For Mandel's problem, we have 40×5 gridblocks, taking the upper domain by symmetry. The sizes of the gridblocks in the x direction are uniform, 0.5 m, while the sizes in the z direction are non-uniform; 0.005 m, 0.01 m, 0.05 m, 0.1 m, 0.1 m. The medium is fully saturated with liquid water, and the initial pressure is 10.0 MPa. We impose drainage boundaries at the left and right sides, and no-flow conditions on the other sides. Initial total stress is also 10.0 MPa over the domain, and we set 20.0 MPa for the overburden, two times greater than the initial total stress. The Young›s modulus and Poisson›s ratio are 450 MPa and 0.0, respectively. The monitoring well is located in the middle of the domain, as shown in Fig. 3. No gravity is considered for the Terzaghi and Mandel problems. The permeability and initial porosity are $6.51\times10^{-15}m^2$, 6.6 mD (1 Darcy=9.87×10^{-13} m^2), and 0.425, respectively for both problems, and the permeability is constant during simulation. Biot's coefficients for both problems are 1.0.

Fig. 4 shows that the staggered (i.e., one iteration) and fully iterated solutions from T+M match the analytical solutions. The numerical results correctly capture the instantaneous pressure buildup for Terzaghi's problem and the Mandel–Cryer effect for Mandel's problem; these could not be captured by a flow-only simulation. From Fig. 5, we can identify fast convergence of the fixed-stress method. When full iteration is taken, the fixed-stress split almost takes 3 iterations for both Terzaghi's and Mandel's problems.

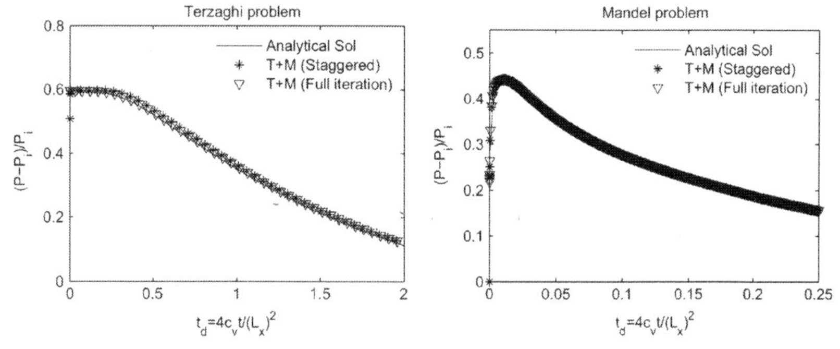

Figure 4: Comparison between numerical solutions of T+M and analytical solutions of Terzaghi's problem (left) and Mandel's problem (right). T+M matches the analytical solutions. cv is the consoli-

dation coefficient, defined as $c_v = k_p(\mu_w(1/K_{dr} + \varphi c_w))$. t is time. Pi is the initial reservoir pressure.

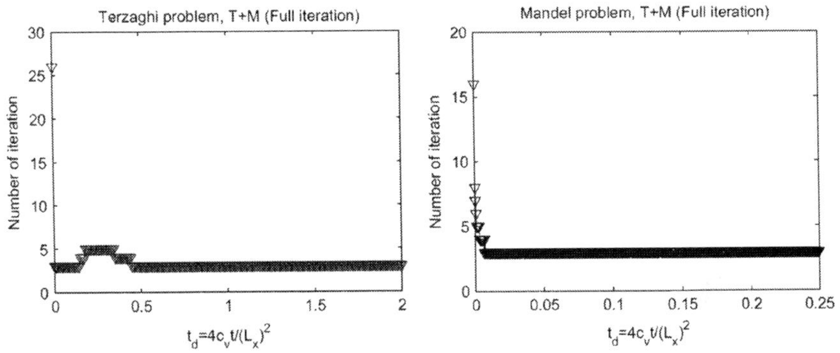

Figure 5: Number of iterations required for convergence with a prescribed tolerance during simulation for T+M when full iteration is performed. The fixed-stress split almost takes 3 iterations for Terzaghi's and Mandel's problems, which shows fast convergence when it takes full iteration. The tolerance of convergence is $|(P^j-P^{j-1})/P_i|_{x=x_0} < 10^{-5}$, where the superscript j indicates the iteration level. x_0 is the monitoring point.

Figs. 4 and 5 support the mathematical and numerical analyses in Kim et al., 2011b and Kim et al., 2013. In the analyses, the staggered and full iterative fixed-stress methods yield numerical convergence in time. In addition, the staggered fixed-stress method yields high accuracy and efficiency. Here, convergence in time means that the numerical solution approaches the true solution when time step size is refined, and accuracy is measured by the error between the numerical solution and the true solution.

1D Elastic Chemo-poro-mechanics

We reuse the domain, boundary conditions, and initial porosity and permeability used in the Terzaghi problem, while having the same initial stress and left boundary stress as the initial fluid pressure. Initially, liquid water is fully saturated, having 8.3 MPa. Young's modulus and Poisson's ratio are 2.4 GPa and 0.2, respectively, and Biot's coefficient is 1.0. We account for reactive transport with chemical reaction, while isothermal single-phase flow is considered. The injection is applied to the right end gridblock at the rate of $1.0 \times 10^{-4} kg^{-1}$. Quartz dissolution is assumed to be the only reaction as observed in the experiment described by Johnson et al. (1998), and modeled using a linear kinetic rate law, written as

$$r_{qz} = -\chi_{qz} A_{qz} \kappa_{qz} \left(1 - \frac{C_{Si}}{C_{Si,qz}}\right), \tag{18}$$

where the subscript $_{(\cdot)qz}$ indicates quartz. Aqz and κqz are the specific surface area and the reaction rate constant, respectively. CSi is the concentration of dissolved (aqueous) quartz, and $_{CSi,qz}$ is CSi in equilibrium. Aqz is $54.2 m^2 kg^{-1}$ (Knauss and Copenhaver, 1995), and κqz is $2.836 \times 10^{-8} molm^{-2} S^{-1}$ at 240 °C (Tester et al., 1994). The initial condition of CSi is zero, which is not in chemical equilibrium at initial time, in order to test the pressure change from the dissolution of quartz. CSi of the injection water is also zero. $_{CSi,qz,0}$ ($_{CSi,qz}$ at initial temperature, 240 °C) is $6.38 \times 10^{-3} mol/kg$. The initial pH of water is 5.589.

Fig. 6 (left) shows comparison of the results between TR-only and T+M for the 1D chemo-poro-mechanics problem. TR-only indicates the TOUGHREACT only simulation. As indicated previously, the coupled flow and geomechanics problem of this test can also be modeled by the flow-only simulation, having $0.9804 \times 10^{-9} Pa^{-1}$ of the exact pore-compressibility, which is calculated from the constrained geomechanics condition of this test case. In this figure, we can see that the results of T+M match the results for TR-only, validating T+M.

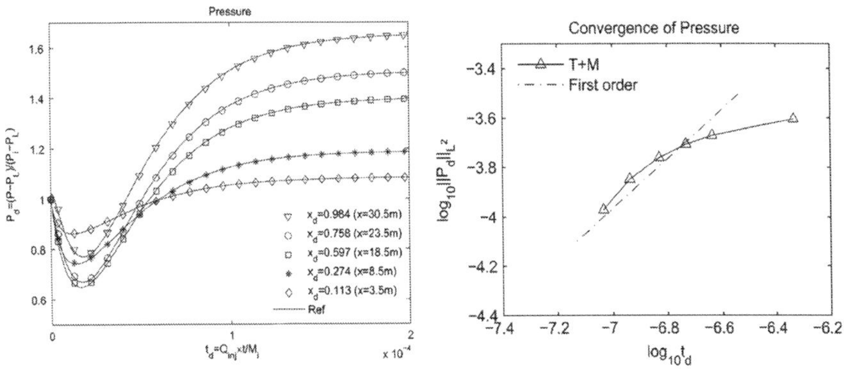

Figure 6: Left: Evolution of pressure and Comparison of pressure between TR-only and T+M for the 1D elastic chemo-poro-mechanics problem. The results from T+M match the results from TR-only. Right: Convergence analysis in time discretization. T+M shows first-order accuracy in time. Mi is the fluid mass in place at initial time, $1.08\times10^4\,kg$. $P_L = 8.2MPa$.

Fig. 7 (left) shows that significant quartz dissolution occurs at early time and at the same rate over the entire domain. This dissolution increases pore-volume and Cqz rapidly (Fig. 7 (right)). The increased pore-volume induces expansion of the pore fluid, liquid water, depending on fluid compressibility. As a result, the fluid pressure decreases over the domain at early time (Fig. 6 (left)).

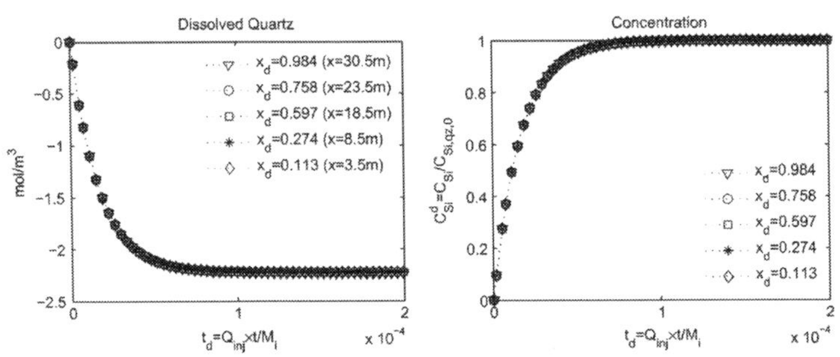

Figure 7: Left: Dissolution of quartz at different monitoring points. Right: Evolution of CSi at different monitoring points. $_{xd}$=x/$_{Lx}$ is the dimension-

less distance, and CdSi is the normalized concentration. As quartz is dissolved into water, CSi increases.

In the left of Fig. 6, the pressure drops are at different points over the domain, because there are influxes to the domain, even though the rates of the dissolution are the same over the entire domain. At the left boundary, liquid water flows into the domain due to the constant pressure boundary, whereas at the right boundary, a constant water-injection rate is applied. Thus, the decrease of pressure at early time is low near the both ends (e.g., xd=0.113 (x=3.5 m) and xd=0.984 (x=30.5 m)), while it is high around the middle of the domain (e.g., xd=0.597 (x=18.5 m) and xd=0.758 (x=23.5 m)).

As CSi increases, as shown in the right of Fig. 7, the rate of dissolution decreases, based on the linear kinetic rate law in Eq. (18). Then pressure increases again because the effect from the influxes becomes more dominant than the dissolution effect. At late time, when the dissolution is tiny, the pressure distribution becomes monotonic without a hump; the highest pressure at the right end (i.e, the injection point), and the lowest pressure at the left end (i.e., the constant pressure boundary).

The right of Fig. 6 shows numerical convergence in time discretization. It should be noted that both TR-only and T+M take the staggered approach. We take a solution of T+M from tiny time step size as a reference solution. From the convergence analysis, we find first order accuracy in time for the staggered T+M. Thus, the sequential algorithm proposed in this paper can increase accuracy by reducing time step size. It should be noted that accuracy cannot be improved even though time step size is reduced, when we use other obvious sequential methods in coupled flow and geomechanics (e.g., non-convergence of the drained method and the fixed-strain method) (Kim et al., 2011a and Kim et al., 2011b).

1D Elastic Chemo-thermo-Poro-mechanics

We further test a 1D elastic non-isothermal chemo-thermo-poro-mechanics, using the same domain, initial and boundary conditions, and the input data as those used in the previous 1D chemo-poro-mechanics problem. Initial temperature is $T_i = 240^\circ C$, and the monitoring point is located at the gridblock of the right end, where liquid water at 0°C is injected at a total mass rate of $Q_{inj} = 1.0\times10^{-2} kgs^{-1}$, in order to investigate noticeable thermal behavior. We have a specific heat capacity of $1000 JKg^{-1\circ}C$, a thermal dilation coefficient of $\alpha_T = 4.5\times10^{-5\circ} C^{-1}$, and a thermal conductivity of $K_T = Wm^{-1\circ}C^{-1}$. In this 1D case, as aforementioned, TR-only can solve the coupled flow and geomechanics problem by using the exact pore-compressibility and thermal expansivity in flow, which are $0.9804\times10^{-9} Pa^{-1}$ and $3.1765\times10^{-4\circ} C^{-1}$, respectively.

Fig. 8 shows comparison of pressure and temperature between 'TR-only' and 'T+M' as well as convergence study of T+M in time discretization. The results of T+M match those of TR-only. T+M shows first-order accuracy in time discretization when time step size is refined, same as the previous 1D elastic chemo-poro-mechanics test. This implies that the proposed sequential method does not lose accuracy of the backward Euler method, which is first-order in time. From the fast convergence of the full iterative approach and the first order accuracy of the staggered approach, the proposed algorithm is comparable to the monolithic (fully coupled) method.

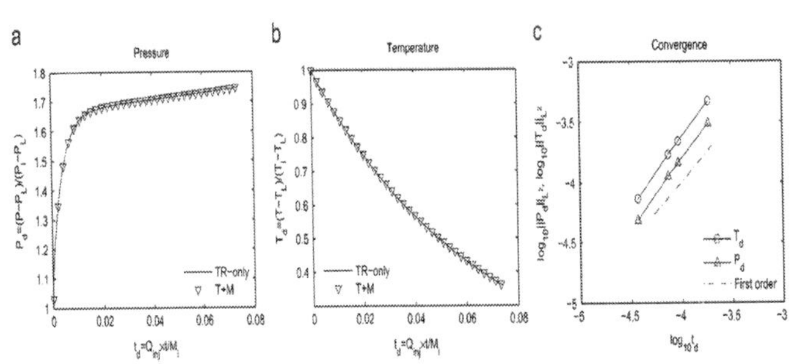

Figure 8: Comparison of (a) pressure and (b) temperature between TR-only and T+M for the 1D elastic chemo-thermo-poro-mechanics problem.

The results from T+M match the results from TR-only. (c) Convergence analysis in time discretization. T+M shows first-order accuracy in time. P_L=0.0Pa, and t_L=0.0°C.

2D Elastic Chemo-thermo-poro-mechanics

We consider a naturally fractured geothermal problem shown in the left of Fig. 9. The geothermal reservoir has a large dimension perpendicular to the fracture plane, i.e., the problem can be reduced to one of multiple porosity 2D plane strain geomechanics, as shown in the right of Fig. 9. For the 2D problem, we have 20×10 gridblocks in the x and z directions, the sizes of which are uniform, 10.0 m. We assume the size of the dimension in the y direction to be 500 m in flow, large enough for the constant temperature boundary for heat flow. Each gridblock has five subelements, one fracture and four rock matrix media, the volume fractions of which are 0.002, 0.098, 0.20, 0.35, and 0.35, respectively.

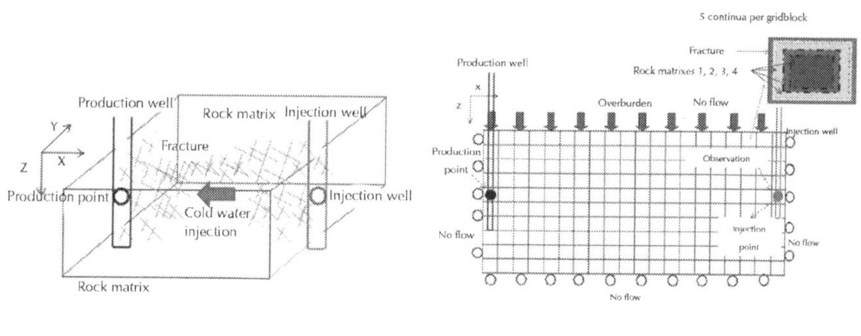

Figure 9: Left: The schematics of a 3D fractured geothermal reservoir. Right: The discretized domain of the fractured geothermal reservoir with the 2D plane strain geomechanics and the multiple porosity model.

The reservoir is fully water saturated, and the initial pressure and temperature at the top are 19.0 MPa and 146.7 °C, respectively, which might be much cooler than a producing geothermal system, but which shows qualitatively the same behavior. They are distributed with the hydrostatic pressure gradient, 9.806 kPa/m, and the geothermal gradient, 25 °C/km. We impose no-flow boundaries of heat and fluid on all sides, assuming an alternative injection and production scenario with symmetry. The monitoring points are the injection and production points at the fracture medium (P1 and P2, respectively), as shown in the left of Fig. 9. For geomechanics, the initial vertical total stress is −19.0 MPa at the top, and vertically distributed based on the bulk density, 2200 kg/m³. The horizontal total stress in the x direction is the same as the vertical stress. The total stress in the y direction is then determined based on the plane strain condition with given geomechanical moduli described below. We have 19.0 MPa overburden, and horizontally constrained boundaries at the sides and no vertical displacement at the bottom. Other flow and geomechanical properties are listed in Table 1.

Table 1: Input data for flow and geomechanics

Permeability	$5.9\times10^{-13}m^2$ (fracture),	$5.9\times10^{-19}m^2$ (rock matrix)
Porosity	0.5 (fracture),	0.3 (rock matrix)
Heat conductivity	$0.5Wm^{-1\circ}C^{-1}$ (desaturated),	$3.1Wm^{-1\circ}C^{-1}$ (saturated)
Specific heat capacity	$100JKg^{-1\circ}C^{-1}$ (all)	
Injection & Production	50.0 kg/s at 20 °C,	50.0 kg/s
Young's modulus	213.75 MPa (fracture),	2.234 GPa (rock matrix)
Poisson's ratio	0.0 (fracture),	0.3 (rock matrix)
Biot's coefficient	1.0 (all)	
Thermal dilation coefficient	$a_T = 4.5\times10^{-5o}C^{-1}$ (all)	

For geochemistry, we take the same quartz dissolution process used in the previous 1D chemo-poro-mechanics problem. We have the initial CSi obtained from the chemical equilibrium at initial time, taking $C_{Si,qz,0} = 2.078\times10^{-3} mol/Kg$ at 147 °C, because we found from the previous 1D chemo-poro-mechanic problem that non-equilibrium at initial time changed pressure and concentration noticeably. It should be noted that $_{CSi,qz}$ is a function of temperature, from which $_{CSi,qz}$ varies during the simulation as temperature changes.

We use dynamic permeability and Young's modulus hereafter, assuming that deformation of fractured medium is small. The assumption of small deformation is consistent with the infinitesimal transformation, which allows us to use Eq. (2). Then, we employ the strain and failure-dependent permeability and the strain-dependent Young's modulus, motivated by the Cubic law (Snow, 1965 and Witherspoon et al., 1980), as follows.

$$k_{p,ii} = k_{p,ii,0}\lambda^{f}(1 + \varepsilon_{ii})^{3}, \quad ii = xx, \; yy, \; zz, \tag{19}$$

where the subscript $_{(\cdot)0}$ indicates the reference state. λf is the permeability multiplier, determined from the failure status. In other words, a high value is assigned dynamically, when a porous medium faces failure. If the material faces shear failure, λ_f becomes $100\times_{Nfg}/_{Ng}$ in this study, where Ng and Nfg are the numbers of total and failed Gauss points with a gridblock, respectively. λf is constant for the 2D elastic problem of this subsection, while it will vary for the 3D elastoplastic problem in the next subsection.

It should be noted that one may use different types of the nonlinear permeability depending on specific reservoir conditions.

Fig. 10 shows the comparison between chemo-thermo-poro-mechanics (with reactive transport) and thermo-poro-mechanics only (without reactive transport). In Fig. 10 (a), for both results, we observe an increase in pressure at early time, resulting from fluid injection as well as thermal contraction. Figs. 10 (b) and (c) show that, as temperature decreases, volumetric strain v decreases, inducing compaction of the reservoir, although fluid is continuously injected. Both volumetric strains with and without reactive transport still decrease even after there is no further decrease of temperature at P1, because compaction at other areas induces the immediate compaction at P1. This compaction results in the additional increases of pressure at P1, as shown in Fig. 10 (a). Then, because of high permeability, pressure at P2 also increases due to the increased pressure at P1, showing the same trend (Fig. 10 (d)). Note that the perturbation of geomechanics at a certain location affects the entire domain instantaneously, because of the nature of quasi-static mechanics. These nonlocal responses, which originate in different time scales between flow and geomechanics, cannot properly be captured by a flow-only simulation, but only by the coupled flow-geomechanics simulation.

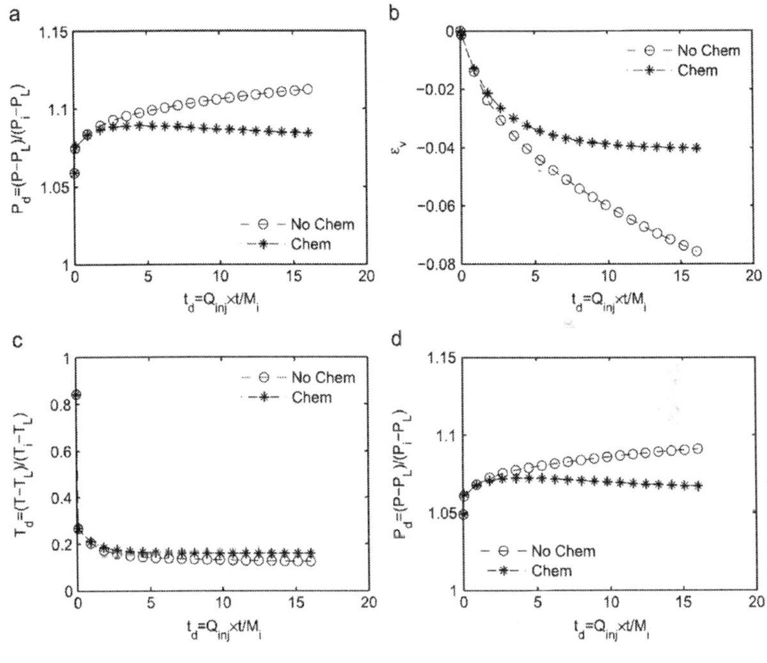

Figure 10: Comparison of the results between chemo-thermo-poro-mechanics, ('Chem'), and thermo-poro-mechanics only, ('No chem'). (a) Pressure at P1, the injection point at the fracture medium. (b) v at P1. (c) temperature at P1. (d) pressure at P2, the production point at the fracture medium. Mi is the initial fluid mass in place at the fracture medium, $1.304\times10^7 Kg$. $P_L = 10MPa$ and $T_L = 0.0^\circ C$.

The comparison between the simulation results with and without reactive transport in Fig. 10 (a) shows that pressure with reactive transport is lower than that without reactive transport, because dissolution of quartz results in an increase of pore-volume, followed by a pressure drop, just as discussed in the previous 1D chemo-poro-mechanics problem. The difference in pressure at P1 causes differences in compaction (Fig. 10 (b)) and pressure at P2 (Fig. 10 (d)).

Fig. 11 shows the evolutions of CSi and the dissolutions of quartz at different locations of the fracture medium (Figs. 11 (a) and (b), respectively), and those at different porous media of the gridblock that contains the injection point (Figs. 11 (c) and (d)). In Fig. 11 (a), CSi is the lowest at the injection point,xd=0.975 (x=195 m) of the fracture medium, because CSi of injected water is zero. Interestingly, as shown in Fig. 11 (b), dissolution of quartz is higher at xd=0.775 (x=155 m) of the fracture medium than at the injection point. Higher temperature at xd=0.775 of the fracture medium than that at the injection point yields higher $_{CSi,qz}$ and higher qz, which in turn lead to higher dissolution rate of quartz. In Fig. 11(c) and (d), CSiwithin the fracture medium is the lowest and dissolution of quartz is the highest within the fracture medium among all the continua of the gridblock, because permeabilities of the rock matrix media are extremely low, and liquid water mainly flows through the fracture medium, not intruding into the rock matrix media.

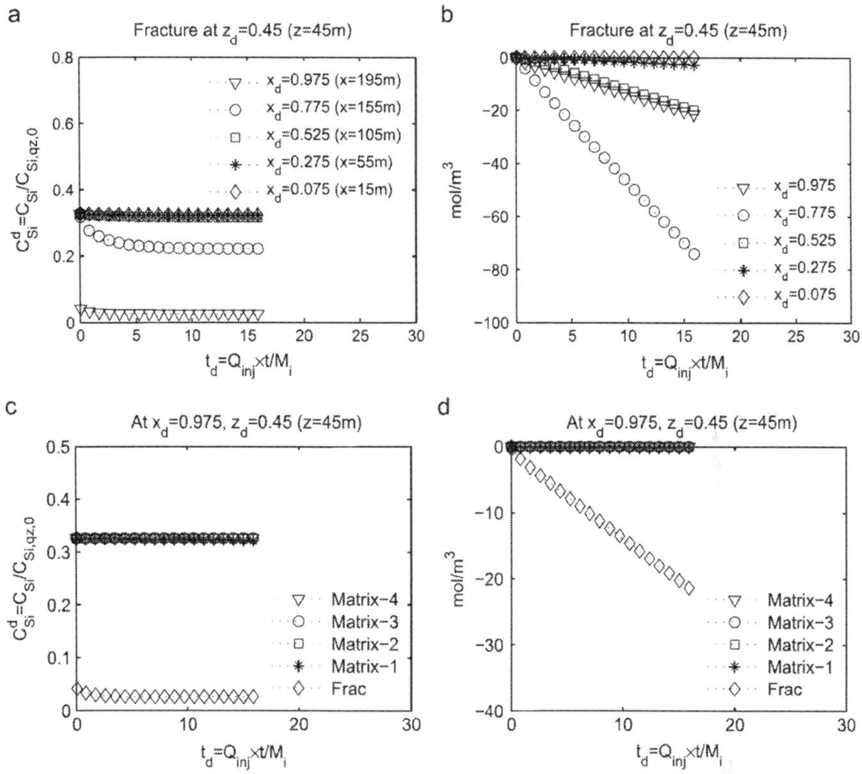

Figure 11: Concentration of aqueous SiO_2, CSi and dissolution of quartz in a chemo-thermo-poro-mechanics simulation. (a) and (c): CdSi at different spatial locations along the fracture medium and CdSi within the fracture medium and rock matrix continua of the injection gridblock. (b) and (d): Dissolution of quartz at the locations corresponding to (a) and (c), respectively. $_{CSi,qz,0}$ is $2.078\times10^{-3} mol/Kg$. 'Frac' implies the fracture medium.

When comparing the coupled flow-geomechanics simulation with the flow-only simulation, we find fromFigs. 10 (a) and 12 significant differences, for example, in pressure and porosity. In the left of Fig. 12, TR-only with reactive transport shows more pressure drop than TR-only without reactive transport, because the dissolution increases pore volume, which causes the pressure to drop below the initial pressure, just as shown in the left of Fig. 6.

On the other hand, we observe that the pressure of chemo-thermo-poro-mechanics is higher than that from TR-only with reactive transport, because the pressure build-up by thermal contraction over the domain completely offsets the pressure drop from the change in pore volume induced by the quartz dissolution.

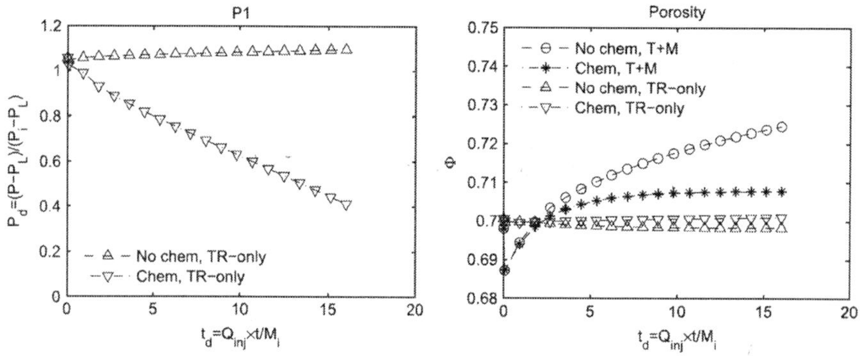

Figure 12: Left: Comparison of the pressure results at P1 between with ('Chem') and without ('No chem') reactive transport in TR-only. Right: Comparison of porosities among chemo-thermo-poro-mechanics, thermo-poro-mechanics only, and TR only with and without reactive transport.

3D Elastoplastic Chemo-thermo-poro-mechanics

We introduce an artificially fracturing geothermal problem. Fig. 13 shows a 3D domain, which has an initial elliptic fracture at the center. We take, by symmetry, the upper and right quadrant as a simulation domain. For the 3D problem, we have 15×15×19 gridblocks in the x, y, and z directions, employing the single porosity model. The sizes of gridblocks in the x and y directions are uniform, 1.0 m, while those in the z direction are not uniform; 1.0 m from the first gridblock to the 8th in depth, 0.2 m, 0.04 m, 0.01 m, 0.04 m, 0.2 m from the 9th to the 13th, respectively, and 1.0 m for the rest gridblocks.

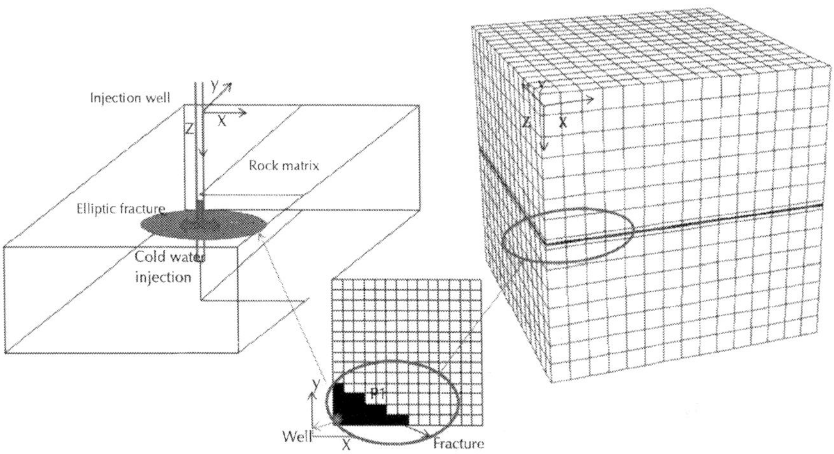

Figure 13: Left: A schematics of a 3D geothermal reservoir with an elliptic fracture. Right: The discretized domain for simulation. The upper and right quadrant is used for the simulation domain.

The reservoir is fully water-saturated, and the initial pressure and temperature at the top are 15.1 MPa and 146.7 °C, respectively. They are distributed with the hydrostatic pressure gradient, 9.806 kPa/m, and the geothermal gradient, 25 °C/km. The initial permeabilities of the fracture and the rock matrix are $5.9\times10^{-13} m^2$ (0.6 Darcy) and $5.9\times10^{-17} m^2$ (0.06 mD), respectively, where we assume the rock matrix to exhibit some microcracks, having higher permeability of the rock matrix than that at the previous 2D problem. We impose no-flow boundaries on all sides. The injection well is located at x=0 m and y=0 m at the 10th layer of the z direction (Fig. 13). The injection rate is 0.25 kg/s with 146.7 °C water, assuming water to become heated at the bottom hole. For geomechanics, the initial vertical total stress is −21.58 MPa at the top, vertically distributed based on the bulk density, 2200 kg/m³. The horizontal total stresses in the x and y directions are both −23.74 MPa at the top and vertically distributed with the gradient of −23.73 kPa/m, which become 1.1 times the vertical total stress. We have 21.58 MPa overburden and no vertical displacement at the bottom. We assign low Young's modulus to the last (15th) horizontal layer in the x and y directions to maintain the initial horizontal stress at the sides. Other input data including geochemistry are the same as those in the previous 2D chemo-thermo-poro-mechanics problem.

For the modeling of plasticity, we consider shear failure from the Mohr–Coulomb model, having $C_h = 10MPa$, and ${}_{f}={}_{d}=\varpi/18$, where f and d are the friction and dilation angles, respectively, and ch is the cohesion. This failure implicitly aims at application of hydro-shearing for the situations where shear failure occurs, activating micro-cracks and stimulating the reservoir.

Once a material faces failure, entering plasticity, a substantial permeability change is applied to simulate the transition from intact rock permeability to that of a fractured rock permeability. Specifically, we dynamically modify the reference permeability of the rock matrix from Eq. (19). We have a monitoring point, P1 (xd=0.5 (x=7.5 m), yd=0.033 (y=0.5 m)), shown in Fig. 13.

Fig. 14 shows comparison of fractured areas with and without reactive transport at the same time, td=9.76 (t=12.1 min). From the comparison, we find differences of the failed areas between the two cases. For the case with reactive transport, we find more extensive horizontal and vertical failures, compared with the case without reactive transport. However, this simulation result does not always mean that simulation with reactive transport cause larger fracturing than simulation without reactive transport, but their differences depend on various reservoir conditions and injection scenarios.

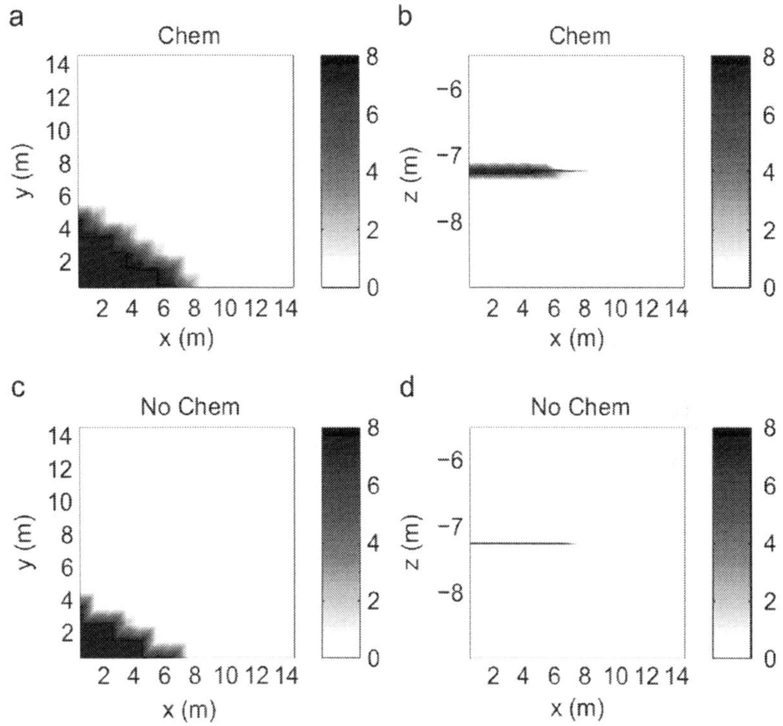

Figure 14: Failure zones at td=9.76 (t=12.1 min) with and without reactive transport at the 10th layer in the z direction ((a) and (c), respectively) and failure zones at the first layer in the y direction ((b) and (d), respectively). Here, $M_i = 74.24Kg$, the initial fluid mass in place at the fracture medium. Values 1–8 indicate the numbers of Gauss points that experience failure, and the initial fractured zone takes 8, while Value 0 indicates no failed zone. Failed areas with reactive transport are larger than those without reactive transport.

Because of this substantial increase of permeability induced by material failure, we find from Fig. 15 that the pressures with and without reactive transport are different at late times. For the case with reactive transport, depressurization due to fast growth of high permeable fractured areas, shown in Fig. 14 and 15(right), seems more dominant than pressurization from injection.

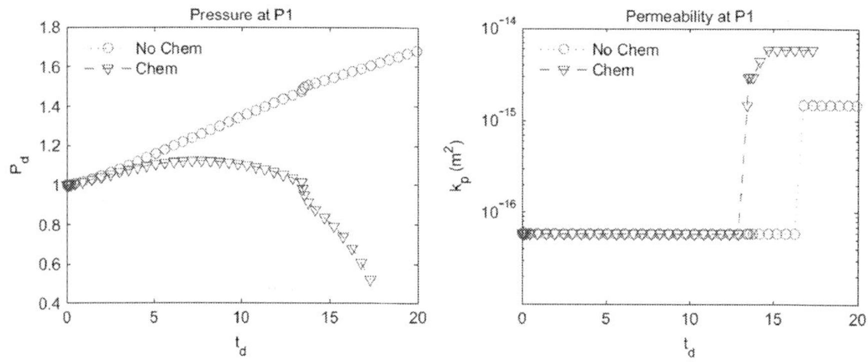

Figure 15: Left: Evolution of pressure (left figure) and permeability (right figure) at P1. Once failure occurs, permeability increases considerably, which affects pressure.

From Fig. 16, CdSi at the injection point decreases considerably from initial time because zero CdSi of water is injected, while CdSi's at other points decrease at a slower rate. Noticeable dissolution of quartz is found as zero CSi of water is injected.

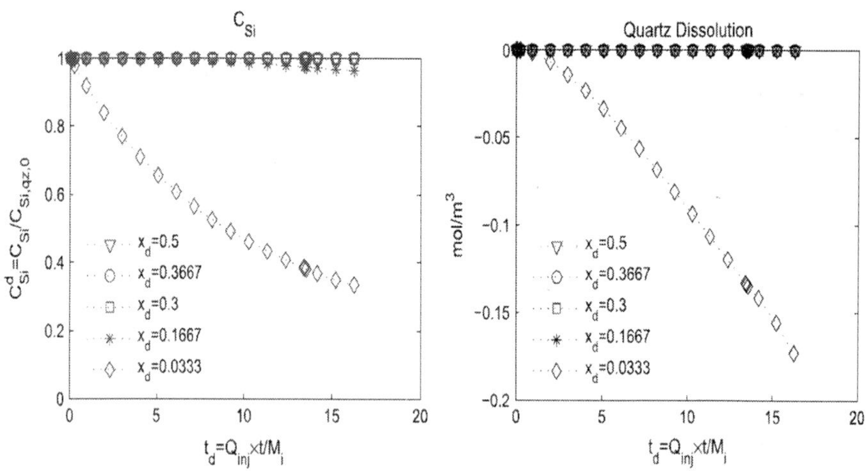

Figure 16: Left: Evolution of CdSi at different locations. Right: Quartz dissolution at different locations. Quartz dissolution occurs as zero CSi of water is injected, and CdSi decreases accordingly.

CONCLUSIONS

We proposed a sequential implicit formulation and algorithm for chemo-thermo-poro-mechanics, and implemented the coupling algorithm within a coupled flow-geomechanics simulator. This simulator can handle the multiple porosity model in fractured geothermal reservoirs, and can consider nonlinear permeability and geomechanics, including plasticity. We validated the coupled geomechanics-flow-reactive transport simulator, using the analytical solutions for the poromechanical effects and the reference solutions that can be computed for the 1D elastic cases. We tested the behavior of numerical convergence for the proposed sequential algorithm, and it yielded fast convergence and kept first-order accuracy in time, comparable to the monolithic approach.

From numerical simulation of the naturally 2D fractured geothermal reservoir with the multiple porosity model, we demonstrated how chemo-thermo-poro-mechanics can yield significantly different results from thermo-poro-mechanics without reactive transport, because small pore volume changes can lead to large fluid pressure effects for chemically sensitive reservoirs in which incompressible fluid is saturated. From the 3D shear fracturing problem induced by fluid injection, we found the differences in failure evolution, causing substantial changes in permeability, for the cases with and without reactive transport. The changes in pore volume, permeability, and fluid pressure caused different evolutions of plastic failure. Thus, we recommend employing an integrated chemo-thermo-poro-mechanics simulator for chemically and thermally sensitive and highly deformable reservoirs such as fractured geothermal reservoirs. The proposed sequential algorithm enables straightforward construction of the interface between existing flow-reactive transport and geomechanics simulators.

ACKNOWLEDGMENTS

This work was supported by the American Recovery and

Reinvestment Act (ARRA), through the Assistant Secretary for Energy Efficiency and Renewable Energy (EERE), Office of Technology Development, Geothermal Technologies Program, of the U.S. Department of Energy under Contract No. DE-AC02-05CH11231, and by 2011 TOUGH2 development grants in the Earth Sciences Division of Lawrence Berkeley National Laboratory.

REFERENCES

1. Armero, F., 1999. Formulation and finite element implementation of a multiplicative model of coupled poro-plasticity at finite strains under fully saturated conditions. Comput. Methods Appl. Mech. Eng. 171, 205–241.
2. Aziz, K., Settari, A., 1979. Petroleum Reservoir Simulation. Elsevier, London.
3. Bai, M., 1999. On equivalence of dual-porosity poroelastic parameters. J. Geophys. Res. 104, 10,461–10,466.
4. Barenblatt, G.E., Zheltov, I.P., Kochina, I.N., 1960. Basic concepts in the theory of seepage of homogeneous liquids in fissured rocks. J,. Appl. Math. 24 (5), 1286–1303.
5. Berryman, J.G., 2002. Extension of poroelastic analysis to double-porosity materials: new technique in microgeomechanics. J. Eng. Mech. ASCE 128 (8), 840–847.
6. Biot, M.A., 1941. General theory of three-dimensional consolidation. J. Appl. Phys. 12, 155–164.
7. Biot, M.A., Willis, D.G., 1957. The elastic coefficients of the theory of consolidation. J. Appl. Phys., 594–601.
8. Coussy, O., 1995. Mechanics of Porous Continua. John Wiley and Sons, Chichester, England.
9. Geertsma, J., 1957. The effect of fluid pressure decline on volumetric change of porous rocks. AIME Petrol. Trans. 210, 331–340.
10. Johnson, J.W., Knauss, K.G., Glassley, W.E., DeLoach, L.D., Tompson, A.F.B., 1998. Reactive transport modeling of plug-

flow reactor experiments: quartz and tuff dissolution at 240 °C. J. Hydrol. 209, 81–111.

11. Kim, J., Moridis, G.J., 2014. Gas flow tightly coupled to elastoplastic geomechanics for tight and shale gas reservoirs: material failure and enhanced permeability. SPE Journal SPE 155640, http://dx.doi.org/10.2118/155640-PA.
12. Kim, J., Sonnenthal, E., Rutqvist, J., 2012. Formulation and sequential numerical algorithms of coupled fluid/heat flow and geomechanics for multiple porosity materials. Int. J. Numer. Methods Eng. 92, 425–456.
13. Kim, J., Tchelepi, H.A., Juanes, R., 2011a. Stability and convergence of sequential methods for coupled flow and geomechanics: drained and undrained splits. Comput. Methods Appl. Mech. Eng. 200, 2094–2116.
14. Kim, J., Tchelepi, H.A., Juanes, R., 2011b. Stability and convergence of sequential methods for coupled flow and geomechanics: fixed-stress and fixed-strain splits. Comput. Methods Appl. Mech. Eng. 200, 1591–1606.
15. Kim, J., Tchelepi, H.A., Juanes, R., 2013. Rigorous coupling of geomechanics and multiphase flow with strong capillarity. SPE J. 18 (6), 1591–1606, Spe 141268.
16. Klinkenberg, L.J., 1941. The permeability of porous media to liquid and gases. API Drill. Produ. Pract., 200–213.
17. Knauss, K.G., Copenhaver, S.A., 1995. The effect of malonate on the dissolution kinetics of albite, quartz and microcline as a function of pH at 70 °C. Appl. Geochem. 10 (1), 14–33.
18. Lewis, R.W., Schrefler, B.A., 1998. The Finite Element Method in the Static and Dynamic Deformation and Consolidation of Porous Media, 2nd edition Wiley, Chichester, England.
19. Moridis, G.J., Kowalsky, M.B., Pruess, K., 2008. TOUGHþHYDRATE v1.0 User›s Manual: a Code for the Simulation of System Behavior in Hydrate-Bearing Geologic Media. Report LBNL-00149E, Lawrence Berkeley National Laboratory, Berkeley, CA.
20. Phillips, P.J., Wheeler, M.F., 2007. A coupling of mixed and

continuous Galerkin fi- nite element methods for poroelasticity I: the continuous in time case. Comput. Geosci. 11, 131–144.

21. Pruess, K., Narasimhan, T.N., 1985. A practical method for modeling fluid and heat flow in fractured porous media. SPE J. 25 (1), 14–26.
22. Pruess, K., Oldenburg, C., Moridis, G., 1999. TOUGH2 User›s Guide. Version 2.0. Report LBNL-43134. Lawrence Berkeley National Laboratory, Berkeley, CA.
23. Rutqvist, J., Stephansson, O., 2003. The role of hydromechanical coupling in fractured rock engineering. Hydrogeol. J. 11, 7–40.
24. Settari, A., Mourits, F., 1998. A coupled reservoir and geomechanical simulation system. SPE J. 3, 219–226.
25. Snow, D.T., 1965. A Parallel Plate Model of Fractured Permeable Media (Ph.D. dissertation). University of California, Berkeley.
26. Sonnenthal, E., Ito, A., Spycher, N., Yui, M., Apps, J., Sugita, Y., Conrad, M., Kawakami, S., 2005. Approaches to modeling coupled thermal, hydrological, and chemical processes in the Drift Scale Heater Test at Yucca Mountain. Int. J. Rock Mech. Min. Sci. 42, 698–719.
27. Taron, J., Elsworth, D., Min, K., 2009. Numerical simulation of thermo-hydrologicmechanical-chemical processes in deformable, fractured porous media. Int. J.
28. Rock Mech. Min. Sci. 46, 842–854.
29. Tester, J.W., Worley, W., Robinson, B.A., Grigsby, C.O., Feerer, J.R., 1994. Correlating quartz dissolution kinetics in pure water from 25 to 625 °C. Geochim. Cosmochim. Acta 58 (11), 2407–2420.
30. Vermeer, P.A., Verruijt, A., 1981. An accuracy condition for consolidation by finite elements. Int. J. Numer. Anal. Methods Geomech. 5, 1–14.
31. Witherspoon, P.A., Wang, J.S.Y., Iwai, K., Gale, J.E., 1980. Validity of Cubic Law for fluid flow in a deformable rock

fracture. Water Resour. Res. 16 (6), 1016–1024.

32. Xu, T., Sonnenthal, E., Spycher, N., Pruess, K., 2006. TOUGHREACT: a simulation program for non-isothermal multiphase reactive geochemical transport in variably saturated geologic media: applications to geothermal injectivity and CO2 geological sequestration. Comput. Geosci. 32, 145–156.

33. Xu, T., Sonnenthal, E., Spycher, N., Pruess, K. 2008. TOUGHREACT User›s guide: a simulation program for non-isothermal multiphase reactive geochemical transport in variably saturated geologic media. LBNL-55460-2008. Lawrence Berkeley National Laboratory, Berkeley, CA.

34. Xu, T., Spycher, N., Sonnenthal, E., Zhang, G., Zheng, L., Pruess, K., 2011. TOUGHREACT Version 2.0: a simulator for subsurface reactive transport under nonisothermal multiphase flow conditions. Comput. Geosci. 37, 763–774.

Chapter 4

Permeability of Sub-critical Carbon Dioxide in Naturally Fractured Indian Bituminous Coal at a Range of Down-hole Stress Conditions

V. Vishal[a, b, c], P.G. Ranjith[b], S.P. Pradhan[c], and T.N. Singh[c]

[a]IITB-Monash Research Academy, Indian Institute of Technology Bombay, Mumbai 400076, India

[b]Department of Civil Engineering, Monash University, Clayton, VIC 3800, Australia

[c]Department of Earth Sciences, Indian Institute of Technology Bombay, Mumbai 400076, India

ABSTRACT

Understanding reservoir dynamics from a geotechnical and petrophysical perspective is essential for safe and steady fluid extraction, transport or injection in the reservoir. The recent approach on storage of anthropogenic greenhouse gases (GHG) for mitigation of climate change in the deep geologic reservoirs has led to fast-paced research on resolving storage related issues. A coal seam is one such target reservoir where CO_2 may be injected for storage; this may be coupled with the enhanced extraction of coalbed methane (CBM). It is known that the interaction of CO_2 with coal induces hydro-mechanical changes; however, detailed investigation is mandatory for site-specific operation; the lack of such studies on Indian coal is the motivation for this research. A naturally fractured coal specimen was chosen and the subcritical CO_2 flow was maintained by monitoring the injection pressures and ambient temperature; the variations in permeability under various scenarios were also investigated. The results indicate that initial CO_2 injection caused coal matrix swelling which took nearly 5–6 h of flow-related-sorption for maximum closure of the natural fractures. Linear variation in flow rate with the pressure gradient across the sample indicated laminar or Darcian flow of CO_2 through the coal fractures. Deeper coal seams have high confinements; the CO_2 permeability results for such in-situ conditions indicated an exponential decline in naturally fractured coal, due to increasing confining pressures. The observed reduction of CO_2permeability with increase in gas inlet pressure is debated and may be attributed fully to matrix swelling or partly to the Klinkenberg effect. This study is vital for successful GHG sequestration in coal seams and for better understanding of the gas flow behaviour in the seams.

INTRODUCTION

Underground storage of greenhouse gases for geologically significant periods of time is being investigated as one of the promising solutions to combat global warming (IPCC, 2005,

Viete and Ranjith, 2007 and Shukla et al., 2010). India, being a developing nation, is a leading producer of greenhouse gases (GHG) and it is projected that the total annual CO_2 emissions in India will increase from 1229 Mt (2005) to 3084 Mt (2030) (Nair et al., 2003). Vishal et al. (2013a) summarised that the potential CO_2 storage in the Indian subcontinent may be targeted in the offshore saline sedimentary basins, depleted oil and gas fields, basalt formations and unmineable coal seams. The recent decade has witnessed a rapid development in the aspects of geological storage of anthropogenic CO_2 (White et al., 2005 and Nasvi et al., 2013). Different basins and horizons from different parts of the world are being evaluated for long term, safe and stable disposal of CO_2. However, very limited investigations are available for the Indian context, and extensive site specific investigations should be carried out.

Coal seam sequestration may be coupled with enhanced recovery of coalbed methane (CBM), which partly offsets the costs of CO_2 capture, transport and storage operations (Jasinge et al., 2011, Hol and Spiers, 2012 and Vishal et al., 2013a). From a geomechanical perspective, the response of coal to CO_2 injection must be well understood. Coal is a naturally fractured, weak geomaterial which experiences changes in structure, strength, and permeability due to adsorption of CO_2 (Krooss et al., 2002, Mehic et al., 2006, Viete and Ranjith, 2006, Viete and Ranjith, 2007, van Bergen et al., 2009, Busch and Gensterblum, 2011, Hol and Spiers, 2012 and Hol et al., 2012). Further, effective stresses, confining stresses and ambient pressure–temperature conditions of target seams control the flow of fluids in the coal strata, primarily (Seidle et al., 1992, Siriwardane et al., 2009 and Vishal et al., 2013b). Coal displays an unconventional behaviour since CO_2 injection involves coal matrix swelling, and the associated closure of fluid pathways leads to reduced permeability. Further, due to high heterogeneity in coal specimens from basin to basin and within the same basin, it is necessary to investigate the changes in the flow characteristics of CO_2 in representative samples from each basin.

UNCONVENTIONAL RESERVOIR CHARACTERISTICS OF COAL

Mechanism of Gas Storage

Compared to conventional oil and gas reservoirs, where the primary fluid storage mechanism is in multiple pore spaces by physical compression, gases in coal are mainly trapped by physico-chemical sorption onto the micropores of solid matrix blocks. In addition, coals have macroporosity which is comprised of the mutually perpendicular cleats that act as channels for the flow of fluids in coal (Law, 1993 and Laubach et al., 1998). Gases in the cleats, known as 'free gases', move mainly due to the pressure gradient (Figure 1). Stored methane molecules diffuse to the cleats from the micropores within the coal matrix, and flow along these to reach the production well. Similarly, the injected CO_2 molecules pass from the injection well and enter the matrix blocks through the cleats and remain trapped in the micropores.

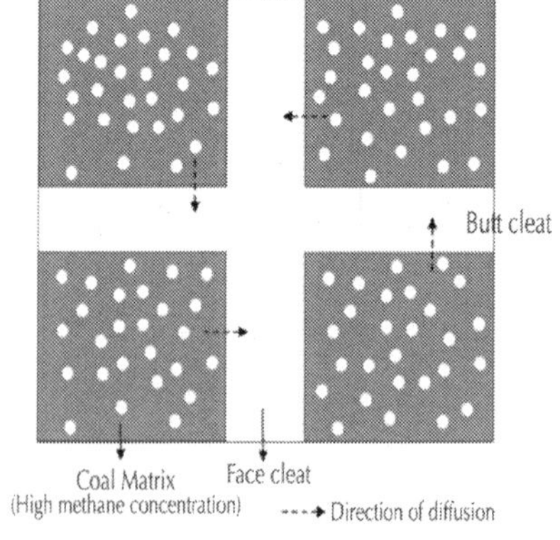

Figure 1: Schematic cross-section of coal structure, representing its dual porosity model.

Adsorption Affinity in Coal

Gases in coal remain trapped by the adsorption mechanism, which depends on pressure, temperature, coal rank, maceral content, coal type, moisture content as well as the gas type. An increase in pressure increases the gas adsorption capacity while the temperature has an inverse effect on coal (Bustin and Clarkson, 1998 and Goodman et al., 2007). High rank coals, mostly with less moisture and high vitrinite content, have higher gas retention potential than the low rank coals (Levy et al., 1997, Crosdale et al., 1998, Krooss et al., 2002, Kelemen and Kwiatek, 2009 and van Bergen et al., 2009). When CO_2 is injected into coal, the residual CH_4 is displaced from the micropores, as coal has a higher sorption affinity towards carbon dioxide over methane. Several studies have established that there is a preference for adsorption of different gases on to the micro-pores of coal; the order of preference, from high to low, is carbon dioxide, methane, hydrogen and nitrogen (Arri et al., 1992, Gentzis, 2000, Mastalerz et al., 2004 and Mazumder et al., 2006). This fact works positively in targeting coal for long term storage of CO_2, with the possibility of enhanced recovery of coalbed methane. It has been shown using numerical simulation that nearly three times more volume of CO_2 may be injected in place of each volume of CBM recovered from the coal seams in one of the Gondwana Basins in India (Vishal et al., 2013a).

Shrinkage/Swelling Behaviour of Coal

During desorption, there is a reduction in the pressure in the coal matrix and pores. This causes a reduction in the volume of the coal matrix, which in turn widens the cleats present in the coal. On the contrary, gas adsorption causes a volumetric increase in the coal matrix. Thus, 'swelling' of coal due to adsorption may be defined as: the increase in the volume occupied by coal, as a consequence of viscoelastic relaxation of its strained, glassy/rubbery and highly cross-linked macromolecular structure (Larsen and Kovac, 1978,Brenner, 1985, Larsen et al., 1997 and Larsen, 2004). It is understood that the formation of new intramolecular bonds between coal and the adsorbate molecules, if thermodynamically favoured, replaces the original coal–coal hydrogen bond. The modification in the coal bonding in presence of adsorbates disrupts the molecular structure that leads to swelling in the coal (Larsen, 2004). The sorption induced volumetric deformation in coal affects the flow rate of gases through coal. Injection of CO_2 induces a higher degree of swelling in a coal matrix, which results in shrinking of gas flow paths, that is, the cleats, thus reducing permeability. The preferred affinity of coal for CO_2 over methane, on one hand, indicates large GHG storage potential and enhanced recovery of methane from coal; however, the higher coal expansion due to CO_2 injection results, significantly, in loss in its permeability and is a major issue in coal seam sequestration (Robertson, 2005). Loss in CO_2 injectivity was observed in several field scale pilot studies.Mavor et al. (2004) reported a nearly four-time reduction in CO_2 permeability in coal in ECBM pilot tests in Alberta, Canada. Botnen et al. (2009) showed a ten-time reduction in CO_2 permeability in lignites of Williston Basin (North Dakota). It was observed that that injection of pure N_2 over some days could reverse the permeability reduction due to CO_2 injection in coal beds of Yubari pilot project, Japan (Shi et al., 2008). However, Perera et al. (2011a) hold that the loss in coal permeability could only be partly reversed due to CO_2 injection.

Coal Permeability in Response to Stress Changes

The arrangement of cleats results in anisotropic permeability in different directions and is one of the most crucial parameters concerning GHG injection in coal. Using water as the fluid medium, Pomeroy and Robinson (1967) demonstrated that flow rate varied significantly with confining stresses in all the principal directions, that is, along the face cleat, butt cleat and bedding planes. Koenig and Stubbs (1986) studied field well test data from the Warrior Basin and found that the permeability anisotropy ratio in two directions of coal was nearly 17:1. Gash et al. (1993) used coal samples from the San Juan Basin and observed that permeability parallel to the butt cleats (0.3–1.0 mD) and that along the face cleats (0.6–1.7 mD) were comparable, while vertical to bedding planes, permeability was two orders lesser (0.007 mD). Migration of fluids in coal fractures/cleats follows either the Darcian or the non-Darcian flow movement, while transport within matrix occurs by diffusion, mostly following Fick's Law.

Fluid injection in conventional hydrocarbon reservoirs is usually associated with an increase in permeability. However, during CO_2 injection in coal, two distinct phenomena with an opposing influence on its permeability take place (Gray, 1987 and Shi and Durucan, 2004). Injection of fluid is associated with an increase in pore pressures in cleats, which reduces the effective stresses in coal; while adsorption of the injected phase causes the coal matrix to swell, reducing cleat spacing, thus increasing the effective stresses. Effective stress in a coal medium may be defined as the confining stresses minus the pore pressure (Harpalani and Chen, 1997, Jaeger et al., 2007 and Ranjith and Perera, 2011). The contradictory phenomenon influences the flow behaviour in coal and is a subject of research, prior to implementation of GHG sequestration. Few studies have focussed mainly on the pure elastic stresses in coal due to fluid pressure in the absence of external stresses in laboratory conditions (Hol and Spiers, 2012). Investigation of stress-induced changes in coal permeability has been carried out by several researchers for coals in different basins (Bustin, 1997, Robertson, 2005 and Jasinge et al., 2011); however, not many details are available for the Indian Gondwana coal seams.

Real Term Implications of CO_2 Permeability in Coal

Injection of CO_2 in coal seams on pilot/demonstration/commercial projects has been initiated and successfully deployed in different countries, mainly for enhanced coalbed methane recovery. However, loss in CO_2 injectivity in most of the cases mainly was attributed to the volumetric expansion of the coal matrix (Ohga et al., 2005, Oudinot et al., 2007 and Botnen et al., 2009). In some cases, however, increase in coal permeability was observed in later phases of CO_2 injection due to reduced effective stresses and reduction in reservoir pressure due to generation of coalbed methane (Perera and Ranjith, 2012).

Hydraulic fractures were created in some cases like the Upper Silesian Basin (Poland) and the Ishikari Basin (Japan) to combat loss in coal permeability and to eventually enhance CO_2 injectivity (Ohga et al., 2005 and Fujioka et al., 2010). Therefore, the role of fractures in coal is extremely important to ensure an easier flow of CO_2 in the coal. The phenomenon of CO_2-induced coal matrix deformation is now significantly established for different coals around the world. Therefore, fractures in coal are important to ensure higher permeability.

It is extremely essential to understand the influence of several critical parameters on the permeability of coal before application in field. These include role of effective stresses, confining stresses, coal matrix swelling, etc. which may be investigated through experiments or numerical simulations. Indian coals are generally of low permeability; this study highlights the CO_2 flow behaviour in fractured bituminous coals and shall lay the foundation of research for long term safe sequestration of CO_2in coal seams.

To date, very little literature on the response of India coals to CO_2 injection is available. Further, a geomechanical and petrophysical perspective of CO_2 injection in coal must be investigated in detail at the in situ P–T conditions of the target coal seams. This will help understand the reservoir response to CO_2injection and the injection pressures with respect to the surrounding stress conditions may be effectively controlled. In this article, we studied the response of naturally fractured Indian coal at different down-hole stress conditions during gas phase CO_2 injection for intermediate depth sequestration.

SAMPLE PREPARATION AND METHODOLOGY

A detailed field study was conducted in the Jharia coalfield in India and block samples were collected from mine sites and transported to the Deep Earth Energy Lab (DEEL) at Monash University (Australia) for testing. The cleat porosity of intact coal samples was as low as 0.28%; therefore, natural fractures in the coal blocks were tapped during coring. The choice of naturally fractured coal specimens was made to enable faster movement of gases through the coal specimen. The objective of this study was to investigate the role of pore pressure, hydrostatic confinements and effective stresses on gaseous CO_2 permeability in coal; hence, the natural fractures in the chosen coal specimens facilitated easier flow of CO_2 due to their higher permeability. The flow of CO_2 through the coal matrix is generally very slow due to larger diffusion paths. Previous studies used fractured coal specimens to investigate the influence of CO_2 exposure on coal permeability through induced fractures (Siriwardane et al., 2009), and to investigate the subcritical and supercritical CO_2 flow in naturally fractured Australian coal (Perera et al., 2011a and Perera et al., 2011b). It is also known that CO_2 changes its phase from gas to liquid or supercritical depending on the pressure and temperature conditions. In this study, high pressures were not reached to investigate for less deep coal seams where CO_2 will be stored in the sub-critical or gaseous state.

Sample Description

Coal formations in India may be divided into two groups, depending on the origin and period of their deposition. Almost all of the coal beds belong to the Permo-Carboniferous (P-C) period, while Tertiary coals account for nearly 1% of the total coal in India. Although most of the studies for CBM exploration have been carried out on P-C coals, some researchers suggest a fairly optimistic sequestration potential in the Tertiary coal units (Chakraborty et al., 2011). P-C coals are located close to large point sources of emissions of CO_2 and are also being exploited for recovery of methane. Thus, a dual-benefit-approach will work for sequestration in P-C coals with or without enhanced CBM recovery. Detailed field studies were carried out on Jharia Basin coalfields to understand the stratigraphic correlation and continuity of the coal seams and to identify locales of coal fire-prone areäs. Jharia Basin is a part of the East–West aligned Damodar–Koel Group of Gondwana Basins of India. The geological map of the Jharia Basin, highlighting the sampling area in the field, is shown in Fig. 2. A brief geological outline of Jharia is mentioned in Table 1(after Chandra, 1992).

Talchir Formation forms the base of the sedimentary succession, unconformably overlying the Achaean basement. The sequence above includes the fluvial and fluvio-lacustrine sediments of the Barakar, Barren Measures and Raniganj Formations (Fox, 1930, Mehta and Murthy, 1957 and Sengupta et al., 1979) deposited within an intracratonic extensional setting (Ghosh and Mukhopadhyay, 1985). There are three lineaments in the Jharia Basin trending NNE–SSW, NE–SW and NW–SE. The southern and northern boundaries of the Jharia coalfield are faulted with several inter- and intra-formational faults. The lithounits dip gently with a mean dip angle of strata equal to 5°. The coal samples used in this study were taken from the Barakar Formation, which consists of medium to coarse sandstone, grey shale overlain by a thin layer of alluvium and thick coal seams. Both, the coal blocks (40 cm by 30 cm) and the coal cores (54 mm diameter), were obtained from the sites and transported to DEEL for estimation of their geomechanical characteristics. The drill hole cores were tested for basic geomechanical characteristics of coal while the blocks were drilled along the natural fractures to get a core sample (of 39 mm diameter) for permeability characterisation under different scenarios of CO_2 injection.

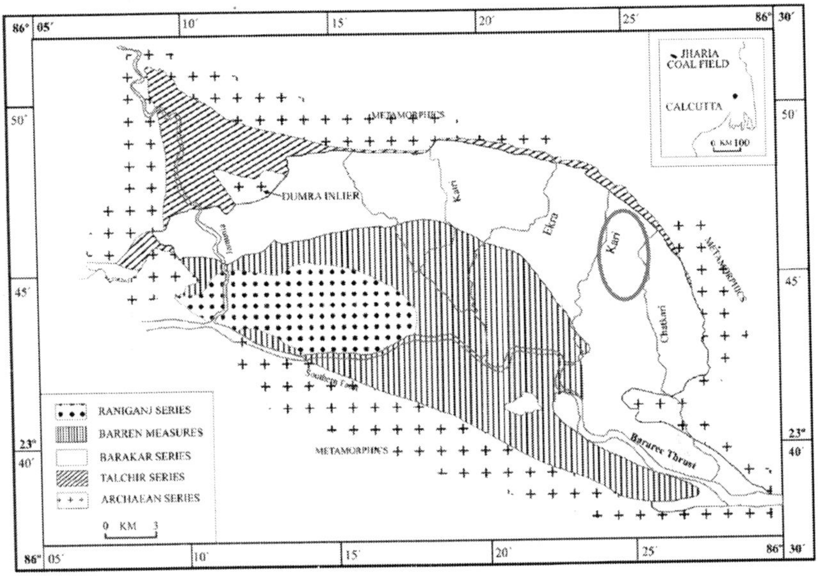

Figure 2: Geological map of Jharia Basin highlighting the sampling area in field (in red oval).After Chandra (1992).

Table 1: General geology of Jharia coalfield

Age	Formation	Dominant lithology
Recent unconformity	Weathered	Sandy soil, clay, alluvium, gravel
Jurassic	Igneous flow and intrusions	Dolerite dykes, lamprophyres, etc.
Upper Permian	Raniganj	Arkosic sandstones, shales and coal seams
Middle Permian	Barren Measure	Shales, carbonaceous shales and buff coloured sandstone
Lower Permian	Barakar	Thick coal seams along with shales, carbonaceous shales, grits and buff coloured sandstones
Upper Carboniferous	Talchir	Greenish shales and fine grained sandstones
Archean	Basement	Metamorphics

After Chandra (1992).

Prior to the permeability experiments, key geomechanical parameters of coal specimens were determined. The coal block had natural fractures; therefore, the strength of coal samples was lesser than that of intact specimens. This helped constraining the experimental conditions, preventing material failure during permeability measurements. A total of 15 samples were tested and the mean strength of the fractured coal was 12.70 ± 0.32 MPa, determined at a load displacement rate of 3 mm/min, while the Young›s modulus was shown to be 4.23 ± 0.14 GPa. Coal is also a low density material as compared to most rocks. The density value was obtained as 1.47 g/cc. The compressional wave velocity of the coal samples was also determined and the average P-wave velocity was found to be 1776 ± 26.17 m/s. A summary of key geomechanical parameters is shown in Table 2. It is clear that the presence of fractures not only reduces the strength of the coal but also results in higher standard deviations in their values due to the random orientation of fractures and the orientation of the sample with respect to the fractures.

Table 2: Pertinent geomechanical characteristics of the naturally fractured coal samples

Rock characteristic	Value	Standard deviation
UCS (MPa)	12.70	0.32
Tensile strength (MPa)	1.93	0.25
Cohesion (MPa)	1.13	0.09
Density (g/cc)	1.47	0.03
P wave velocity (m/s)	1776.3	26.17
Slake durability index (%)	91.72	1.31
Young's modulus (GPa)	4.23	0.14
Poisson's ratio	0.237	0.033
Point load index	0.373	0.016

Apparatus and Procedures

The apparatus and testing procedure have been described by Shukla et al. (2012), though the original set-up was appropriately modified in accordance with the objectives of study of coal (Vishal et al., 2013b). A schematic diagram of the experimental set-up is shown in Fig. 3a. The experimental set-up mainly comprises of four units: (i) loading unit having the loading frame and pressure cell, (ii) plumbing system made up of confining unit, gas injection unit, gas outlet unit, (iii) temperature control unit and (iv) data measurement system. The sample was placed inside the cell in between the top and bottom pedestals as shown in Fig. 3b. Silicone rubber gel was applied on all possible openings to prevent leakage of gas from the core. The assembly was fitted onto the base of a triaxial cell and all lines for fluid flow were tested well for any leakage. The cell was filled with hydraulic oil and a constant confining pressure was applied. The temperature was maintained at a constant 26 °C to prevent any temperature-related change in properties of the confining oil as well as the injected CO_2.

During experiments, CO_2 was injected through the bottom ports at the lower end of the sample. Pressure transducers and gauge valves attached to each pressure line facilitated data acquisition and manual pressure checking of upstream and downstream lines of the specimen. The pressure transient method was used to analyse the flow rate of the flowing medium since the original flow meter could record flow rates only up to 50 mL/h. The downstream pressure was monitored and recorded at regular time intervals using the data acquisition system.

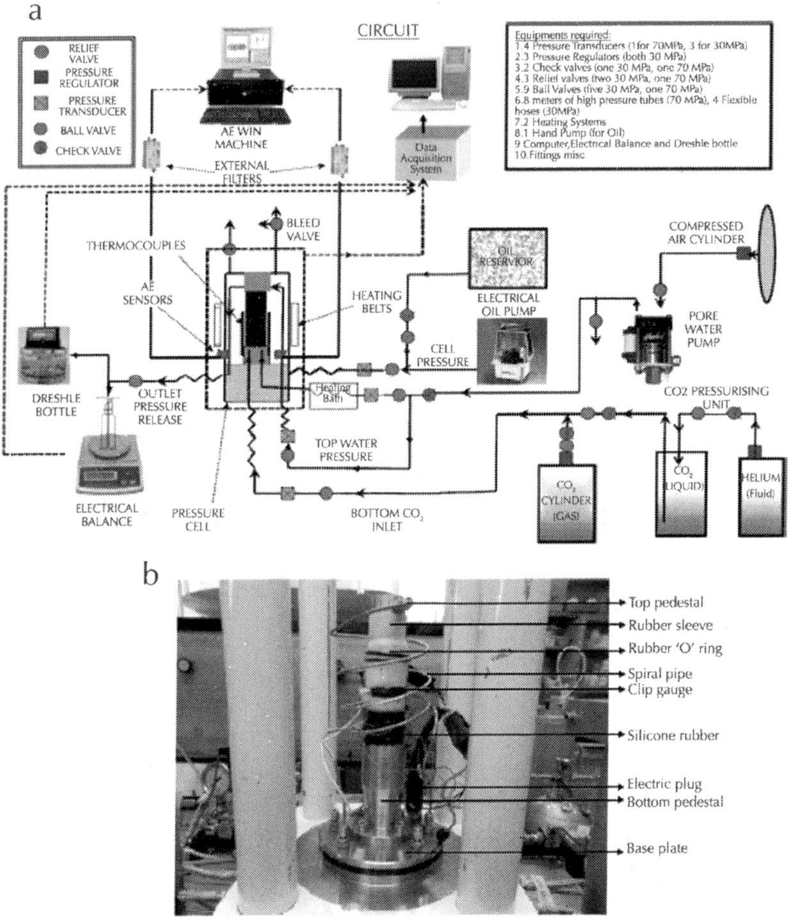

Figure 3: a: Schematic diagram of the pressure–temperature controlled triaxial set-up for permeability testing (From Shukla et al., 2012). b: Close up view of a sample put in a rubber sleeve and sealed using silicone rubber gel; placed in between the top and base pedestals, mounted onto the cell base.

The changes in pressure were obtained and used to calculate the flow-rate using the following equation (Siriwardane et al., 2009 and Ranjith and Perera, 2011):

$$Q = \frac{dP}{dt}\beta V \tag{1}$$

where, Q is the downstream flow rate, dP/dt is the rate of change in downstream pressure with time, V is the downstream volume (evaluated by filling water in the downstream area and eliminating the water load by applying a head) and β is the adiabatic compressibility of the fluid phase. The corresponding permeability of the coal specimens under each condition for CO_2 injection was calculated using Darcy's Law:

$$K = \frac{2Qp_o\mu L}{A(p_i^2 - p_o^2)} \tag{2}$$

where, Q is the downstream flow rate, μ is the viscosity of the injecting fluid, p_o is the downstream pressure, p_i is the upstream pressure, and L and A are the mean length and the cross-sectional area of the coal specimen, respectively. Experiments were performed at a confining pressure (Pc) of 5, 7, 9, 11 and 13 MPa, with CO_2 injection pressures (Pis) being increased stepwise (0.5 MPa steps) from 1 MPa to 5 MPa. The exception was the 5 MPa Pc experiment for which the maximum CO_2 pressure was 3.5 MPa to maintain Pc > Pi. The experiments were done on a single sample, and by continuously increasing injection pressure and confining pressure.

RESULTS AND DISCUSSIONS

The main objective of this study was to investigate the role of confining pressures, pore pressures and effective stresses on permeability of coal to sub-critical CO_2, where presence of natural fractures facilitated easier flow of CO_2. CO_2 was injected into the coal core sample at different injection pressures and the development of downstream pressure was constantly monitored. The rate of increase of pressure at the downstream was used to estimate the CO_2 flow rate through the coal specimen. Three minutes of flow of CO_2 was determined to be the constant time for ensuring a complete steady state flow of CO_2 through the coal sample in each case. The flow rate was estimated using Eq. (1) with the gas pressure gradient, the adiabatic compressibility and dynamic viscosity of CO_2 at the respective experimental pressure and temperature. The latter parameters were determined using the REFPROP database, which uses equations of thermodynamics and the transport properties of a fluid to calculate the state points of a fluid or mixture.

Effect of Confining Pressure (Pc) On Flow Rate of CO_2 in Fractured Coal

Our experiments showed that the flow rate of CO_2 increased with increasing injection pressure (Pi). This was primarily due to the increase in fluid pore pressure and reduction in effective stresses due to increased Pi, which caused possible pore dilation and therefore, enhanced gas flow rate. Although it was expected that CO_2 would lead to volumetric expansion of coal, no decline in flow rate was observed with increasing injection pressure. Further, since the increase in gas flow rate was linear to the pressure gradient, the flow was assumed to be laminar; therefore, Darcy's Law was applied for estimation of permeability for each state of gas injection. CO_2 flow rate in coal may be linear or nonlinear and hence, Darcy's and non-Darcy's Laws have previously been applied by earlier researchers for correct estimation of coal permeability.Jasinge et al. (2011) successfully demonstrated the difference in estimated coal permeability using Darcy's and non-Darcy's equations. They indicated that the variation of flow rate with injection pressure must be monitored accurately for correct estimation of coal permeability. Perera et al., 2011b and Perera et al., 2012 also obtained a linear variation of the flow rate with coal injection pressure, while Jasinge et al. (2011) observed a non-linear variation; therefore, they used different laws for correct determination of coal permeability. The results obtained in this study, as plotted in Fig. 4, clearly indicate linear variation. The trend lines added to the data set at each confining pressure also show good alignment of data points along the line. High values of coefficient of determination (R^2) ranging from 0.959 to 0.995 were obtained for straight line fits; therefore, the flow of CO_2 along the coal specimen was inferred to be Darcian flow. The calculations also revealed a strongly negative relation of the flow rate with the confining pressure (Figure 5). This was primarily due to closure of the cleat system in coal with a higher confining pressure. A good correlation was obtained for a negative exponential relation between the CO_2 flow rate and the confining pressure. The exponential relation gave a higher coefficient of determination than that for a linear or logarithmic fit. For Pi = 1 MPa, the CO_2 flow rate reduced by nearly 75% as the confining pressure was raised from 5 MPa to 7 MPa. This includes the influence of coal confinement as well as coal swelling during the first phase of CO_2 injection.

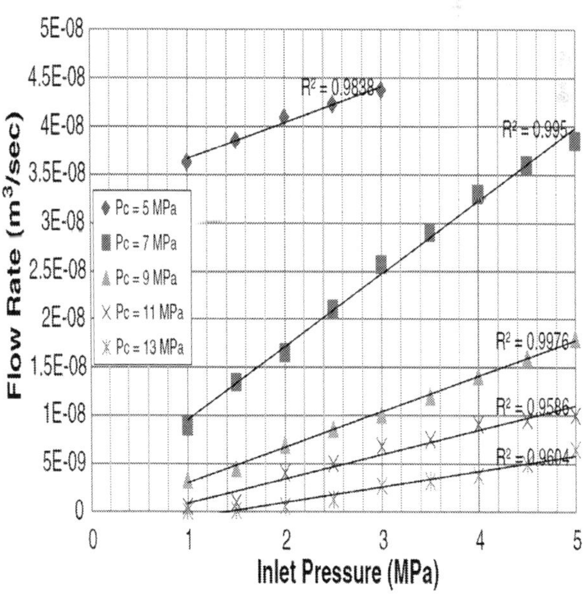

Figure 4: Variation of flow rate vs. inlet pressure is linear which enabled application of Darcy's Law for estimation of permeability.

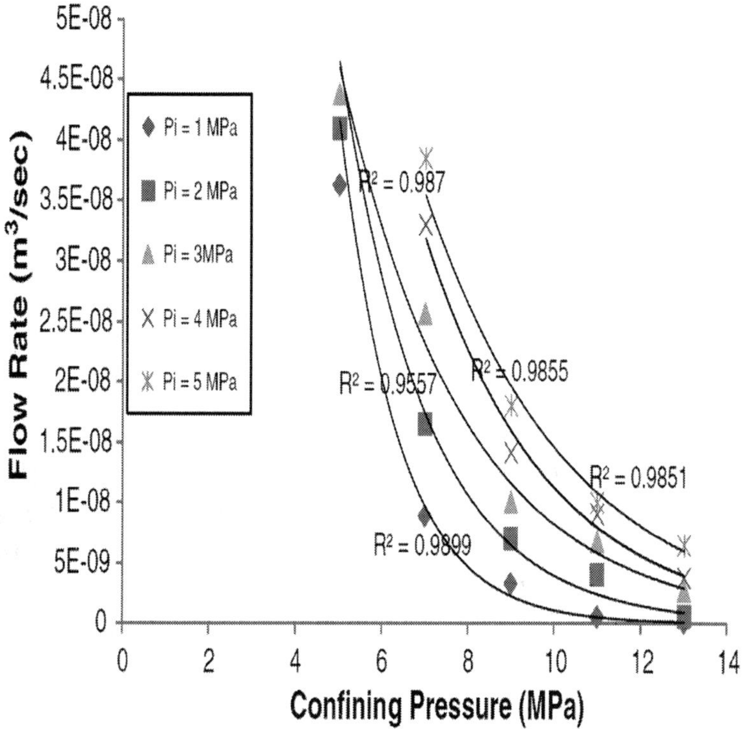

Figure 5: Variation of CO_2 flow rate with confining pressure.

Effect of Confining Pressure (Pc) On CO_2 Permeability in Fractured Coal

With increasing confining pressures, that is, depth, the natural fractures in coal get closed, resulting in the less permeability values obtained for deep seated coal seams. The role of confinement on CO_2permeability in coal is plotted in Fig. 6. Each sub-figure shows the variation of permeability with every 0.5 MPa increment in the gas inlet pressure. It can be observed that the permeability of CO_2 in a naturally fractured coal specimen decreased with the injection pressure (even though the flow rate increased at low Pi (< 2.5 MPa)), while it became almost constant with Pi at higher injection pressures. At a constant Pc of 5 MPa, coal permeability to CO_2 reduced by nearly 32% from first to second case of CO_2 injection. The rate of this permeability reduction with increasing Pi though decreased at high confining pressures (reduced by nearly 23% at 13 MPa confinement) and finally, the permeability of coal to CO_2 increased linearly with gas injection pressure at 13 MPa confinement (Figure 6e). The decrease in permeability with Pi may be attributed to the progressive coal matrix expansion, which closes the path of gas flow. Though CO_2 flow stabilised very fast, the process of matrix swelling expectedly continued for a longer duration; hence, in most of the cases, reduction in permeability with increasing inlet pressure was observed. Minor dilations of fractures due to high injection pressures were compensated by the closure of coal fractures due to swelling and no significant change in permeability was observed. Due to the consistent exposure of CO_2, most of the expansion would have taken place during the early four confinements (when Pc varied from 5 MPa to 11 MPa); hence, increase in permeability with increasing inlet pressure was observed at 13 MPa confinement pressure. Once the matrix related swelling effect fades away, an increase in the injection pressure leads to dilation of pores, resulting in increase in permeability. Such arguments also support the results of Perera et al. (2011c) who observed that the naturally fractured Australian coal specimen experienced maximum coal matrix swelling during the first 3–4 h of sub-critical and super-critical CO_2injection. Similarly, Siriwardane et al. (2009) established that the exposure time for stability in case of Pittsburgh coal samples with induced fractures was nearly 20–30 h. Such a long exposure time (nearly 5–6 h) to CO_2 injection in the tested coal samples would have also led to significant swelling of coal on either side of the fractures leading to reduction in CO_2 permeability. The plot of CO_2 permeability with varying confining pressures was constructed, at the common pore pressure of 3.0 MPa (Figure 7). The pore pressure was determined by taking the gas injection pressure and downstream pressure into account at the time of measurement. An exponential decrease in permeability with increase in hydrostatic confinement was observed. The experimental results showed that CO_2 permeability reduced from 0.031 mD at 5 MPa to 0.0001 mD at 13 MPa confining pressure for the same initial injection pressure of 1.0 MPa (a reduction of more than two orders of magnitude). This trend is inferred to be due to the closure of large pores or fractures in coal due to high confinements. This highlights the real implication of deep seam sequestration—low permeability of deeper coal beds and further loss in CO_2 injectivity are among the biggest challenges in the area of coal seam sequestration.

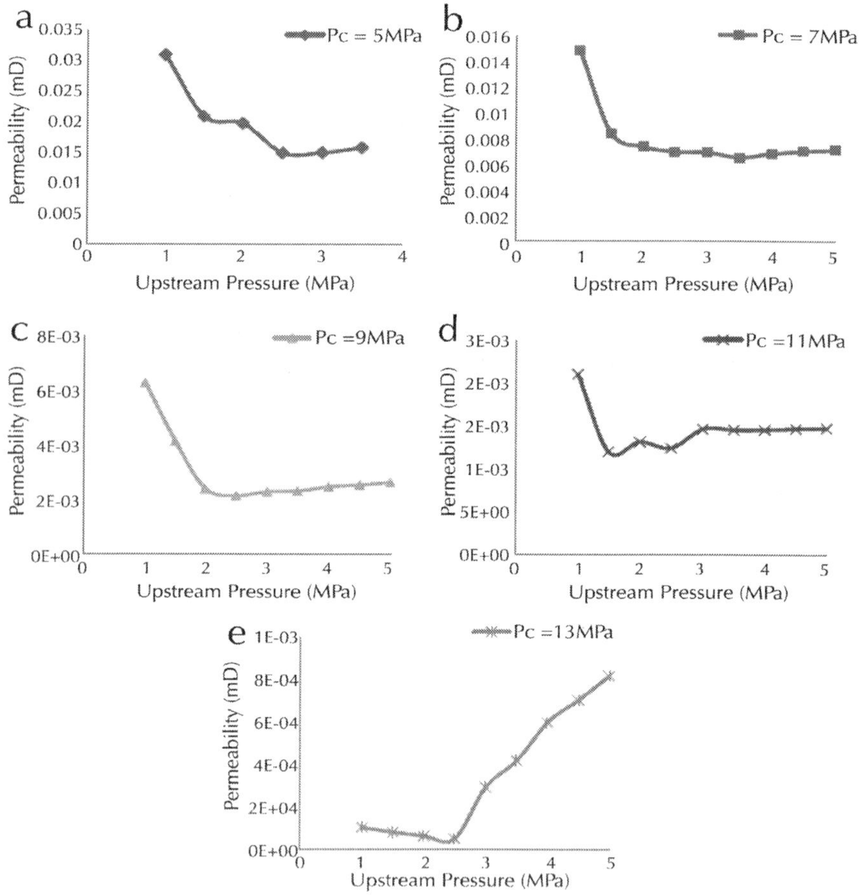

Figure 6: (a–e) Represents variations in permeability of coal with change in confining pressure from 5 MPa to 13 MPa at an interval of 2 MPa.

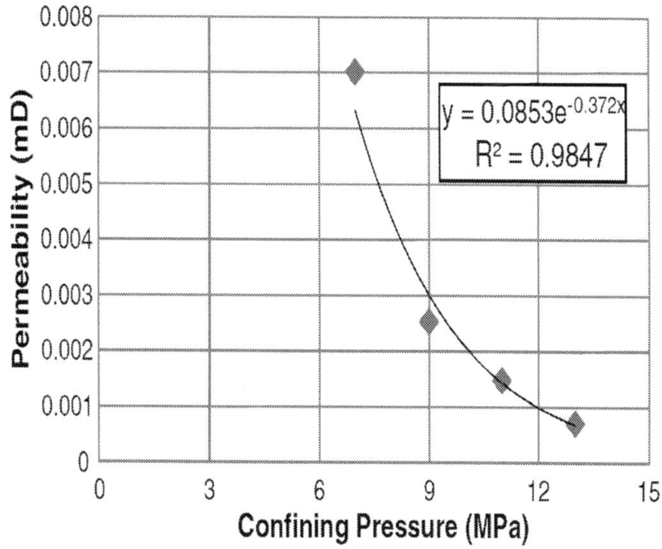

Figure 7: Permeability changes as a function of confining pressure, as can be seen here for a constant pore pressure of 3 MPa.

Reduction in Coal Permeability with Increment in Gas Injection Pressure: Swelling or Klinkenberg Effects?

The reduction in permeability of coal may be attributed to the role of coal matrix expansion. CO_2 gets preferentially adsorbed onto the surface of the coal. Adsorption-induced swelling begins as soon as coal is exposed to the gas. CO_2 gets adsorbed on the fractured surfaces of coal almost instantly, after which it migrates to internal pore spaces by diffusion. This adsorption leads to coal matrix swelling and in turn, closure of the fractured network. The closure of internal fractures due to progressive swelling during flow of CO_2 leads to reduced permeability, when, most of CO_2 transport in the sample occurs by diffusion, through the coal matrix. As compared to the behaviour of an intact coal, the first four cases (when Pc varied from 5 MPa to 11 MPa) were marked by a sharp decline in permeability due to gas injection. Primarily, this happened because the fracture spacing was not completely closed and the movement of gas caused swelling which in turn led to a decline in coal permeability. Finally, the closure of fractures, at the highest confining pressure led to a flow with increasing permeability at higher injection pressures. This was due to opening of the fracture by increasing the fracture aperture which enabled easier flow. Hence, one theory for reduction in coal permeability is dominantly supported by the coal matrix swelling upon CO_2 injection.

The other hypothesis for reduction in coal permeability with increasing inlet pressure relates to the 'Klinkenberg effect' which is often associated with the fluid flow in porous media. Klinkenberg (1941) established that when the pore radius reaches the mean free path of gas molecules, the frequency of the collisions of the gas molecules with the solid matrix walls increases, which leads to increase flow of gas. Earlier researchers have maintained that a higher flow of the subcritical or gas phase of CO_2 at a low injection pressure in the coal as well as other geomaterials may be related to this effect (Kranzz et al., 1979,Abbas et al., 1999, Perera et al., 2011b, Perera et al., 2011c, Takla et al., 2011 and Nasvi et al., 2012). The reduction in coal permeability despite increase in gas injection pressure may partly or fully be contributed by swelling, while a part of higher initial permeability may be due to the Klinkenberg effect.

Effect of Effective Stress on Coal Permeability

With increasing depths, the confining and overburden pressures increase. In addition to the other pressures acting on the coal seam, the fluid injection pressure is also important to ensure steady injection. The fluid pore pressure in coal evolves continuously due to injection of CO_2; the effective stresses on coal samples change accordingly. Effective stress is a function of the confining pressure and the pore pressure. The pore pressure in each case was the average of the injection (upstream) pressure and the pressure developed at the end of specimen (downstream pressure) at the time of measurement of permeability (3 min after the start of injection). Effective stress also has a coefficient of deformation involved, which is assumed to be unity, in this study. A general trend of exponential decline in permeability with increasing effective stresses was observed, while clusters of outliers were also identified. During initial exposure of CO_2 at low injection pressure, coal matrix swelling was dominant leading to high permeability at low injection pressure. This phase of high permeability at higher effective stresses is represented by the clusters highlighted in red in Fig. 8. During passage of CO_2 through the coal sample, the gas may be adsorbed even during transportation and this process is time-dependent (Fokker and van de Meer, 2004). This volume is called as the sorbed volume and is responsible for causing significant adsorption induced volumetric deformation. The permeability of coal reduced from an average of 0.02 mD at low effective stresses to a value of nearly 0.00012 mD at high effective stresses, a reduction by magnitude of order of two. At effective stress close to 12 MPa, the permeability of coal to sub-critical CO2 reduces to 7.71×10^{-5} mD. This provides an impetus to maintain higher injection pressures (which lowers the effective stresses) for efficient CO_2 injection during an actual field operation.

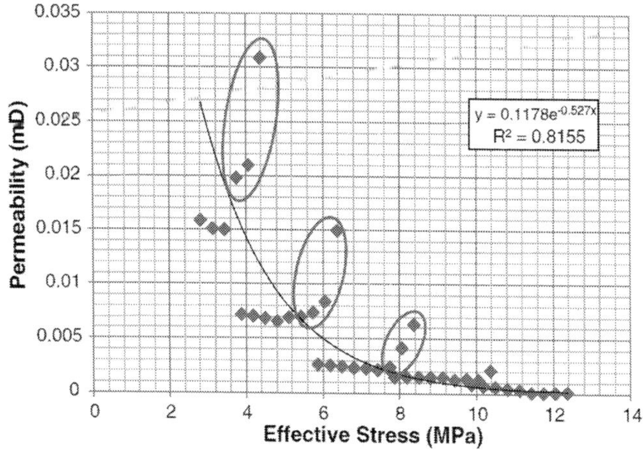

Figure 8: Permeability of fractured coal declines exponentially with effective stresses.

CONCLUSIONS

The recent advances in geo-sequestration of anthropogenic CO_2 to mitigate increasing concentrations has led to coal seams being opted for as target sink spaces for greenhouse gases. Very limited research is available for Indian coalfields, which are among the world's richest coal reserves. The real term implications of coal seam sequestration need to be understood in detail for specific sites because of the extremely heterogeneous nature of coal. Therefore, this study was carried out to understand the consequences of CO_2 injection on the permeability of Indian bituminous coal at different confining and injection pressures.

Linear variation of the flow rate with injection pressure was observed, which suggests Darcian flow of injected CO_2 along the coal sample. Flow rate reduced by nearly 75% during increase in confining pressure from 5 MPa to 7 MPa. It was found that coal seam permeability significantly reduced with increasing confining pressures; a reduction of more than 90% when the confining pressure was raised from 5 MPa to 13 MPa. Apart from creating higher effective stresses, increased confinement on coal expectedly closed the internal fractures leading to less connected conduits for passage of gases, causing reduced permeability. Therefore, a higher pore pressure would be required to maintain higher permeability; low effective stresses during initial periods of confinement caused a higher pore pressure of the fluid leading to easier flow of CO_2. At the same time, an increase in CO_2 injection pressure also reduced permeability by nearly 20–35%, which may partly or fully be accounted for by the coal matrix swelling. Furthermore, higher permeability at low gas injection pressures may have been due to the Klinkenberg effect. These parameters are crucial for ensuring an uninterrupted injection of CO_2 in coal seams. The challenges in the processes need to be identified and addressed so that India is prepared to commit to geological sequestration of readily increasing anthropogenic CO_2 emissions.

REFERENCES

1. Abbas, A., Carcasses, M., Ollivier, J.P., 1999. Gas permeability of concrete in relation to its degree of saturation. Mater. Struct. 32, 3–8.
2. Arri, L.E., Yee, D., Morgan, W.D., Jeansonne, M.W., 1992. Modelling Coalbed Methane Production with Binary Gas Sorption. SPE Rocky Mount. Reg. Meet., Casper Wyoming, SPE-24363.
3. Botnen, L.S., Fisher, D.W., Dobroskok, A.A., Bratton, T.R., Greaves, K.H., McLendon, T.R., et al., 2009. Field test of CO2 injection and storage in lignite coal seam in North Dakota. Energy Procedia 1 (1), 2013–2019.
4. Brenner, D., 1985. The macromolecular nature of bituminous coal. Fuel 64, 167–173.
5. Busch, A., Gensterblum, Y., 2011. CBM and CO2-ECBM related sorption processes in coal: a review. Int. J. Coal Geol. 87 (2), 49–71.
6. Bustin, R.M., 1997. Importance of fabric and composition on the stress sensitivity of permeability in some coals, northern Sydney Basin, Australia: relevance to coalbed methane exploration. AAPG Bull. 81, 1894–1908.
7. Bustin, R.M., Clarkson, C.R., 1998. Geological controls on coalbed methane reservoir capacity and gas content. Int. J. Coal Geol. 38, 3–26.
8. Chakraborty, A., Tiwari, P.K., Singh, A.K., 2011. Coalbed methane exploration in a tertiary lignite basin, North Gujarat, India. GEO India, 2nd South Asian Geosc Conf, New Delhi.
9. Chandra, D., 1992. Jharia coalfield. Mineral Resources of India. Geol Soc India 149.
10. Crosdale, P.J., Beamish, B.B., Valix, M., 1998. Coalbed methane sorption related to coal composition. Int. J. Coal Geol. 35 (1–4), 147–158.
11. Fokker, P.A., van de Meer, L.G.H., 2004. The injectivity of

coalbed CO2 injection wells. Energy 29 (9–10), 1423–1429.

12. Fox, C.S., 1930. The Jharia coalfield. Mem. Geol. Surv. India 56, 255.
13. Fujioka, M., Yamaguchi, S., Nako, M., 2010. CO2-ECBM field tests in the Ishikari coal basin of Japan. Int. J. Coal Geol. 82, 287–298.
14. Gash, B.W., Volz, R.F., Potter, G., Corgan, J.M., 1993. The effects of cleat orientation and confining pressure on cleat porosity, permeability and relative permeability in coal. The 1993 Int Coalbed Methane Symp. Univ Alabama, Tuscaloosa, Alabama, USA. Paper 9321.
15. Gentzis, T., 2000. Subsurface sequestration of carbon dioxide — an overview from an Alberta (Canada) perspective. Int. J. Coal Geol. 43, 287–305.
16. Ghosh, S.K., Mukhopadhyay, A., 1985. Tectonic history of the Jharia Basin — an intracratonic Gondwana basin of eastern India. Q. J. Geol. Min. Metall. Soc. India 57, 33–58.
17. Goodman, A.L., Busch, A., Bustin, R.M., Chikatamarla, L., Day, S., Duffy, G.J., et al., 2007. Intlaboratory comparison II: CO2 isotherms measured on moisture equilibrated Argonne premium coals at 55 °C and up to 15 MPa. Int. J. Coal Geol. 72 (3–4), 153–164.
18. Gray, I., 1987. Reservoir engineering in coal seams: part 1 — the physical process of gas storage and movement in coal seams. SPE Reserv. Eng. 2 (1), 28–34 (SPE-12514-PA).
19. Harpalani, S., Chen, G., 1997. Influence of gas production induced volumetric strain on permeability of coal. Geotech. Geol. Eng. 15 (4), 303–325.
20. Hol, S., Spiers, C.J., 2012. Competition between adsorption-induced swelling and elastic compression of coal at CO2 pressures up to 100 MPa. J. Mech. Phys. Solids 60 (11), 1862–1882.
21. Hol, S., Spiers, C.J., Peach, C.J., 2012. Microfracturing of coal due to interaction with CO2 under unconfined conditions. Fuel 97, 569–584.

22. IPCC, 2005. IPCC Special Report on Carbon Dioxide Capture and Storage, Prepared by the Working Group III of the Intergovernmental Panel on Climate Change. In: Metz, B., Davidson, O., de Conninck, H.C., Loos, M., Meyer, L.A. (Eds.), Cambridge Univ Press, Cambridge, UK and New York, USA, p. 442.
23. Jaeger, J.C., Cook, N.G.W., Zimmerman, R.W., 2007. Fundamentals of Rock Mechanics. Blackwell Publishing, Malden.
24. Jasinge, D., Ranjith, P.G., Choi, S., 2011. Effects of effective stress changes on permeability of Latrobe Valley brown coal. Fuel 90, 1292–1300.
25. Kelemen, S.R., Kwiatek, L.M., 2009. Physical properties of selected block Argonne Premium bituminous coal related to CO2, CH4 and N2 adsorption. Int. J. Coal Geol. 77 (1–2), 2–9.
26. Klinkenberg, L.J., 1941. The permeability of porous media to liquids and gases. Drill Product Pract. 200–213.
27. Koenig, R.A., Stubbs, P.B., 1986. Interference testing of a coalbed methane reservoir. The SPE Unconventional Gas Technology Symposium. Soc Petrol Eng. Richardson, Texas, USA (SPE 15225).
28. Kranzz, R., Franke, A., Engelder, T., Scholz, C., 1979. The permeability of whole and jointed barre granite. Int. J. Rock Mech. Min. Sci. Geomech. Abstr. 16, 225–234.
29. Krooss, B.M., Bergen, F., Gensterblum, Y., Siemons, N., Pagnier, H.J.M., David, P., 2002. High-pressure methane and carbon dioxide adsorption on dry and moisture equilibrated Pennsylvanian coals. Int. J. Coal Geol. 51, 69–92.
30. Larsen, J.W., 2004. The effects of dissolved CO2 on coal structure and properties. Int. J. Coal Geol. 57, 63–70.
31. Larsen, J.W., Kovac, J., 1978. Polymer structure of bituminous coals. ACS Symp. Ser. 71, 36–49.
32. Larsen, J.W., Flowers II, R.A., Hall, P., Carlson, G., 1997. Structural rearrangement of strained coals. Energy Fuel 11,

998–1002.

33. Laubach, S.E., Marrett, R.A., Olson, J.E., Scott, R.A., 1998. Characteristics and origin of cleat: a review. Int. J. Coal Geol. 35, 175–207.
34. Law, B.E., 1993. The Relation between Coal Rank and Cleat Spacing: Implications for the Prediction of Permeability in Coal. Proc Int Coalbed Methane Symp II, pp. 435–442.
35. Levy, J., Day, S.J., Killingley, J.S., 1997. Methane capacity of Bowen Basin coals related to coal properties. Fuel 74, 1–7.
36. Mastalerz, M., Gluskoter, H., Rupp, J., 2004. Carbon dioxide and methane sorption in high volatile bituminous coals from Indiana, USA. Int. J. Coal Geol. 60 (1), 43–55.
37. Mavor, M.J., Gunter, W.D., Robinson, J.R., 2004. Alberta multiwell micro-pilot testing for CBM properties, enhanced methane recovery and CO2 storage potential. 2004 SPE Annual Tech Conf Exhib, Houston, Texas, 26–29 September, Paper SPE 90256.
38. Mazumder, S., van Hemert, P., Busch, A., Wolf, K.H.A., Tejera-Cuesta, P., 2006. Flue gas and pure CO2 sorption properties of coal: a comparative study. Int. J. Coal Geol. 67, 267–279.
39. Mehic, M., Ranjith, P.G., Choi, S.K., Haque, A., 2006. The Geomechanical Behavior of Australian Black Coal under the Effects of CO2 Injection: Uniaxial Testing. Geotech Special Pub Am Soc Civil Eng, Reston VA 20191-4400, United States, pp. 290–297.
40. Mehta, D.R.S., Murthy, B.R.N., 1957. A revision of the geology and coal resources of the Jharia Coalfield. Mem. Geol. Surv. India 84 (2), 142.
41. Nair, R., Shukla, P.R., Kapshe, M.M., Garg, A., Rana, A., 2003. Analysis of long-term energy and carbon emission scenarios for India. Mitig. Adapt. Strateg. Glob. Chang. 8 (1), 53–69.
42. Nasvi, M.C.M., Ranjith, P.G., Sanjayan, J., 2012. The permeability of geopolymer at downhole stress conditions: application for carbon dioxide sequestration wells. Appl. Energy 102, 1391–1398.

43. Nasvi, M.C.M., Ranjith, P.G., Sanjayan, J., Haque, A., 2013. Sub- and super-critical carbon dioxide permeability of wellbore materials under geological sequestration conditions: an experimental study. Energy 54, 231–239.
44. Ohga, K., Fujioka, M., Yamaguchi, S., 2005. Pilot test of CO2 sequestration and ECBM in
45. Japan: reduction of CO2 emission by means of CO2 storage in coal seams. Presented at Silesian Coal Basin of Poland Workshop. Szczyrk, Poland.
46. Oudinot, A.Y., Schepers, K.C., Reeves, S.R., 2007. Gas injection and breakthrough trends as observed in ECBM sequestration pilot projects and field demonstrations. Presented at Coal Bed Methane Symp, Paper 0714, Alabama.
47. Perera, M.S.A., Ranjith, P.G., 2012. Carbon dioxide sequestration effects on coal›s hydro-mechanical properties: a review. Int. J. Energy Res. 36, 1015–1031.
48. Perera, M.S.A., Ranjith, P.G., Airey, D., Choi, S.K., 2011a. Sub-critical and super-critical carbon dioxide flow behaviour in naturally fractured black coal: an experimental study. Fuel 90 (11), 3390–3397.
49. Perera, M.S.A., Ranjith, P.G., Choi, S.K., Airey, D., 2011b. Investigation of temperature effect on permeability of naturally fractured black coal: an experimental and numerical study. Fuel 94, 596–605.
50. Perera, M.S.A., Ranjith, P.G., Choi, S.K., Airey, D., 2011c. The effects of sub-critical and super-critical carbon dioxide adsorption-induced coal matrix swelling on the permeability of naturally fractured black coal. Energy 36 (11), 6442–6450.
51. Perera, M.S.A., Ranjith, P.G., Viete, D., Choi, S.K., 2012. Effective factors for the performance of coal mass natural cleat system upon CO2 sequestration process in deep coal seams: experimental and numerical study. Int. J. Coal Geol. http://dx.doi.org/10.1016/ j.coal.2012.03.010.
52. Pomeroy, C.D., Robinson, D.J., 1967. The effect of applied stresses on the permeability of a middle rank coal to water.

Int. J. Rock Mech. Min. Sci. 4, 329–343.

53. Ranjith, P.G., Perera, M.S.A., 2011. A new triaxial apparatus to study the mechanical and fluid flow aspects of carbon dioxide sequestration in geological formations. Fuel 90 (8), 2751–2759.
54. Robertson, E.P., 2005. Measurement and modeling of sorption-induced strain and permeability changes in coal. Idaho Nat Lab (INL/EXT-06-11832).
55. Seidle, J.P., Jeansonne, M.W., Erickson, D.J., 1992. Application of matchstick geometry to stress dependent permeability in coals. Proc. SPE Rocky Mount. Reg. Meet, Casper, WY, May 18–21, SPE-24361, pp. 432–444.
56. Sengupta, N., Guha, P.K.S., Mukhopadhyay, A., 1979. Pattern of lower Gondwana sedimentation, Jharia Basin, a model. IVth Int Gond Symp II, Hindustan Publ, Delhi, India, pp. 617–625.
57. Shi, J.Q., Durucan, S., 2004. Drawdown induced changes in permeability of coalbeds: a new interpretation of the reservoir response to primary recovery. Transp. Porous Media 56, 1–16.
58. Shi, J.Q., Durucan, S., Fujioka, M., 2008. A reservoir simulation study of CO2 injection and
59. N2 flooding at the Ishikari coalfield CO2 storage pilot project, Japan. Int. J. Greenhouse Gas Control 2 (1), 47–57.
60. Shukla, R., Ranjith, P., Haque, A., Choi, X., 2010. A review of studies on CO2 sequestration and caprock integrity. Fuel 89 (10), 2651–2664.
61. Shukla, R., Ranjith, P.G., Haque, A., Choi, X., 2012. A novel testing apparatus for hydromechanical investigation of rocks in carbon geosequestration. Rock Mech. Rock Eng. http://dx.doi.org/10.1007/00603-012-0241-2.
62. Siriwardane, H., Haljasmaa, I., McLendon, R., Irdi, G., Soong, Y., Bromhal, G., 2009. Influence of carbon dioxide on coal permeability determined by pressure transient methods. Int. J. Coal Geol. 77 (1–2), 109–118.

63. Takla, I., Burlion, N., Shao, J.F., Saint-marc, J., Garnier, A., 2011. Effects of the storage of CO2 on multiaxial mechanical and hydraulic behaviors of oil-well cement. J. Mater. Civ. Eng. 23, 741–746.
64. van Bergen, F., Spiers, C., Floor, G., Bots, P., 2009. Strain development in unconfined coals exposed to CO2, CH4 and Ar: effect of moisture. Int. J. Coal Geol. 77 (1–2), 43–53.
65. Viete, D.R., Ranjith, P.G., 2006. The effect of CO2 on the geomechanical and permeability behaviour of brown coal: implications for coal seam CO2 sequestration. Int. J. Coal Geol. 66 (3), 204–216.
66. Viete, D.R., Ranjith, P.G., 2007. The mechanical behaviour of coal with respect to CO2 sequestration in deep coal seams. Fuel 86 (17–18), 2667–2671.
67. Vishal, V., Singh, L., Pradhan, S.P., Singh, T.N., Ranjith, P.G., 2013a. Analog modeling of Gondwana coal seams in India as coalbed methane reservoirs substituted for carbon dioxide sequestration. Energy 49, 384–394.
68. Vishal, V., Ranjith, P.G., Singh, T.N., 2013b. CO2 permeability of Indian bituminous coals: implications for carbon sequestration. Int. J. Coal Geol. 105, 36–47.
69. White, C.M., Smith, D.H., Jones, K.L., Goodman, A.L., Jikich, S.A., LaCount, R.B., et al., 2005.
70. Sequestration of carbon dioxide in coal with enhanced coalbed methane recovery: a review. Energy Fuels 19 (3), 659–724

Chapter 5

Contribution of the Exploration of Deep Crystalline Fractured Reservoir of Soultz to the Knowledge of Enhanced Geothermal Systems (EGS)

Albert Genter[a], Keith Evans[b], Nicolas Cuenot[a], Daniel Fritsch[a], and Bernard Sanjuan[c]

[a]GEIE exploitation minière de la chaleur, route de Soultz, BP 40038, 67250 Kutzenhausen, France

[b]ETH Zurich, Engineering Geology, Department of Earth Science, Swiss Federal Institute of Technology, 8092 Zürich, Switzerland

[c]BRGM, Département géothermie, 3, avenue C.-Guillemin, BP 36009, 45060 Orléans cedex 2, France

ABSTRACT

Over the past 20 years, the Soultz experimental geothermal site in Alsace, France, has been explored in detail by the drilling of five boreholes, three of which extend to 5 km depth. Data on geology, fluid geochemistry, temperature, microseismicity, hydraulics and geomechanics have been collected and interpreted by the various teams from the participating European countries and their international collaborators. Two reservoirs have been developed within granite at depths of 3.5 and 5 km. The reservoir at 3.5 km was formed from two wells, 450 m apart, both of which were subjected to hydraulic stimulation injections.

The system was circulated continuously for 4 months at 25 kg/s in 1997 using a downhole pump, and yielded results that were extremely encouraging. The impedance reduced to 0.1 MPa/l/s, the first time this long-standing target had been attained. Construction of a deeper system began shortly afterwards with the drilling of 3 deviated wells to 5 km true vertical depth, where the temperature was 200 °C. The wells were drilled in a line, 600 m apart at reservoir depth, and all were hydraulically stimulated and subjected to acidization injections.

The 3-well system was circulated under buoyancy drive for 5 months in 2005 with injection in the central well, GPK-3, and production from the two outer wells, GPK-2 and GPK-4. This showed good linkage between one doublet pair, but not the other. Further acidization operations on the low-productivity well led to its productivity increasing to almost the same level as the other wells. Construction of a power plant at the site was completed in 2008 and a trial circulation with a production pump in one well and the other shut-in was conducted with power production. Downhole pumps are now installed in both production wells in preparation for long-term circulation of the system. In this article we present an overview of the principal accomplishments at Soultz over the past two decades, and highlight the main results, issues identified, and lessons learnt.

INTRODUCTION

EGS in the World

Enhanced (or engineered) geothermal systems (EGS), previously known as hot dry rock (HDR) systems, arose from a concept initiated in Los Alamos (USA) and Cornwall (UK) for exploiting the vast energy resource that resides as heat in the low-permeability rocks which underlie most continental regions at practically-drillable depths. The central concept is to engineer hydraulic linkages between two or more boreholes within the target reservoir to permit circulation of fluid through the hot rock at rates of commercial interest. A great deal of work has been conducted in classic HDR settings to learn how to engineer linkages that have the favourable characteristics of a heat exchanger (see MIT report on the future of geothermal energy (Tester et al., 2006)). The Soultz reservoir significantly deviates from a classic HDR system in as much as it contains permeable structures that host substantial volumes of natural brines. Over 20 years of intense exploration and testing of the Soultz reservoir has produced a huge quantity of geoscientific data that is valuable not only to geothermal development, but to the Earth science community at large.

The EGS Challenges

Soultz-like EGS projects are characterized by the occurrence of large volumes of natural brines contained in fractured crystalline rocks, whose natural permeability is very low due to the poor hydraulic connection within the fracture network. To achieve circulation through the rock mass at rates needed for commercial energy production, the permeability of the rock mass must be increased or "stimulated" by applying various techniques such as hydraulic or chemical stimulation. Hydraulic stimulation acts mechanically by shear dilation through the injection of large-volumes of water at high pressure, but generates nuisances such as induced seismicity. Chemical stimulation dissolves the secondary minerals sealing the natural fractures and thus can also increase the productivity. However, mineral precipitation in the fractures due to water-rock interactions can also decrease the permeability. If the stimulation effect is successful, geothermal wells are then better connected to a far field reservoir which can be exploited by hydraulic circulation. Thus, hot natural brine is pumped from the production well, transfers its energy to the power plant at surface to produce electricity and/or heat and is then re-injected at a lower temperature into the fractured rock mass.

The development of EGS technology by deep drilling within the upper crustal lithosphere faces many scientific and technical challenges. Among them, a major issue is to improve poor hydraulic connections between production/re-injection wells and the fractured reservoirs by optimizing well trajectories, taking into account of stress field orientation and fracture geometry. Moreover, the physical processes and mechanics of induced seismicity in geothermal systems are still poorly understood and represent a major challenge to be addressed. Finally, the long term hydraulic circulations between wells inducing flow, heat transfer and chemical reactions must be evaluated and their behaviour modelled and forecasted.

The Soultz Adventure

The Soultz geothermal project was initiated by a French-German team in 1986 (Gérard and Kappelmeyer, 1987 and Gérard et al., 1984) and has thus been running for more than 20 years. Early phases of the project were supported entirely by public funds from France, Germany, the European Union and a contribution from Switzerland, but since 1996 the work has been increasingly co-funded by the energy industry. The project has produced a substantial body of data that has proven relevant not only to geothermal but also to the geosciences community as a whole. The drilling of three boreholes to 5 km, the deepest penetration of crystalline rock in France, has yielded fundamental insights into the geology, characteristics of natural fractures, fluid geochemistry, temperature and hydraulic properties of deep crystalline rock masses. However, the immediate focus of the various European teams involved in the project (Baumgaertner et al., 1998, Baria et al., 1999 and Gérard et al., 2006) has been the development, testing and modelling of two EGS reservoirs within the basement at 3.5 and 5 km depth. A rich and diverse set of data has been collected, including geological (hydrothermal mineralogy, structural geology, petrophysics), hydraulic, borehole logging, microseismic and active seismic data, that provides insight into the hydro-thermo-mechanical and geochemical behaviour of a deep crystalline basement subject to forced fluid flow conditions. These 20 years of scientific and technical activities have produced about 40 PhD theses and more than 200 publications in peer reviewed journals. This article presents an overview of the main phases of the project, and highlights the key results and milestones that mark the progress in this on-going project to date.

PHASES OF THE SOULTZ PROJECT

Work at the Soultz site can be broken into three phases: a preparatory phase, a drilling, exploration and reservoir development phase that extended from 1987 to 2007, and a power plant construction phase. The preparatory phase involved the compilation of existing literature and a site survey conducted using data from existing (oil industry) holes and the reinterpretation of old petroleum reservoir measurements. The drilling, exploration and reservoir development phase consisted of three sequential campaigns (Table 1). After an early phase of exploration drilling (GPK-1, EPS-1) at shallow depth (2 km), GPK-1 was deepened to 3600 m and GPK-2 drilled to 3880m. The Soultz project then obtained convincing results between 1991 and 1997. A 4 month circulation test was successfully achieved between these 2 wells in the upper fractured granite reservoir at 3.5 km. Based on these encouraging results, GPK-2 was recased and deepened, and two further deviated wells (GPK-3 and GPK-4) were drilled down to 5 km depth between 1999 and 2004 to reach down-hole temperatures of 200 °C (Fig. 1). The geothermal wells are fully cased from surface down to 4.5 km depth and are in open section below with an 8.5 inches diameter. Geothermal water is pumped from the production wells (GPK-2, GPK-4) and re-injected at lower temperature into the injection well GPK-3. In plan, the 3 deep-deviated wells are roughly aligned with the N170°E orientation which is the main orientation of both the fracture network and that of the present-day maximum principal horizontal stress (Fig. 2).

Table 1: Main phases of the Soultz project between 1984 and 2009

1984–1987	1987–1991	1991–1998	1999–2007	2007–2008
Preparatory phase	Exploration phase	Creation of the two wells system GPK1/GPK2 at –3600 m	Creation of the three wells system GPK2/ GPK3/GPK4 at –5000 m	Construction of the first power production unit
Literature compilation	Drilling GPK1 at –2000 m	Deepening of GPK1 at –3600 m and hydraulic stimulation	Deepening of GPK2 at –5080 m and hydraulic stimulation	Power plant construction
Seismic survey reprocessing and interpretation	Coring EPS1 at –2227 m	Drilling of GPK2 at –3880 m and hydraulic stimulation	Drilling of GPK3 at –5100 m and hydraulic stimulation	Installation of the LSP pump in GPK2 at –250 m
Permitting and drilling preparation		Circulation test between the 2 wells (4 months)	Drilling of GPK4 at –5270 m and hydraulic stimulation	Inauguration of the power plant mid 2008
			Circulation test between the 3 wells (5 months)	Installation of the ESP pump in GPK4 at –500 m
			Complementary chemical stimulations	Inter wells circulation tests

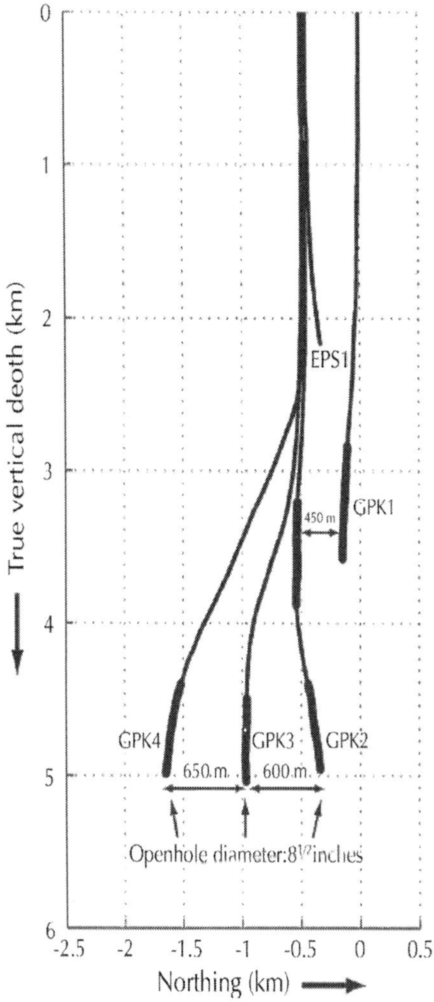

Figure 1: South-north vertical cross-section through the Soultz site showing the location of the upper and lower Reservoirs. The view shows the three 5 km deep geothermal wells (GPK-2, GPK-3, GPK-4) that are completed open-hole in the lower reservoir, and the wells GPK-1 and GPK-2 that formed the upper reservoir. The thicker lines indicate the open-hole sections of the geothermal wells. Also shown is the cored exploration well EPS-1. Depths are expressed in True Vertical Depths (TVD).

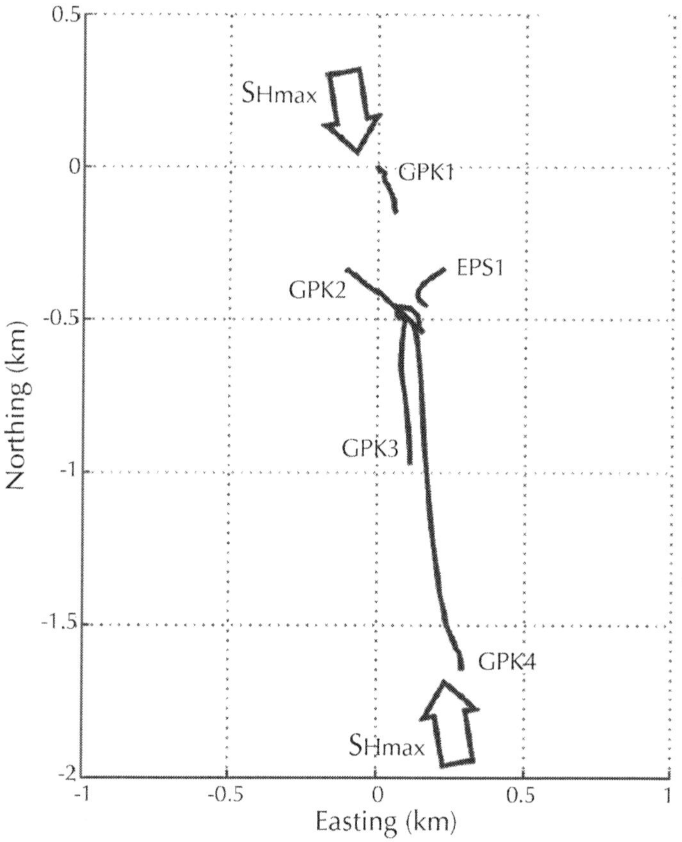

Figure 2: Local map view of the Soultz site, GPK-1 being a local reference. The trajectories of the deep geothermal wells GPK-2, GPK-3 and GPK4 are roughly parallel to the maximum horizontal stress SHmax. GPK-1 and EPS-1 are former exploration wells.

The main results and learning curve deduced over the 20 years of the Soultz basement exploration and exploitation can be summarized as follows. The temperature gradient between 2 and 3.5 km was not as high as expected due to the occurrence of natural circulation within fractures. Hence, to achieve the industry's requirement of 200 °C, the boreholes were deepened to 5 km. Geothermal wells were drilled from the same pad in order to simplify the layout of surface equipment. However, although it was challenging to drill deviated boreholes in hard, fractured, crystalline rocks at 5 km depth, the times to completion in each case were only 5–8 months. Drilling and testing experience showed that not all fractures were productive, reflecting the hydraulic complexity of the natural fracture system with its attendant hydrothermal alteration. Finally, hydraulic circulation tests conducted at various depths showed that hydraulic connections between wells up to 600 m apart that can be established are rather good (Fig. 1). However, it is clear that a major contribution to the hydraulic linkage comes from the pre-existing, far-field structures that constitute a natural reservoir.

THE SOULTZ GEOTHERMAL SYSTEM FROM REGIONAL TO CONCESSION SCALE

The Soultz geothermal area is located in the Upper Rhine Graben (URG), which is part of the European Cenozoic rift system that extends from the Mediterranean to the North Sea coast (Ziegler, 1992). The URG started to form during Late Eocene in response to NNE trending Alpine compression (Bourgeois et al., 2007) and in the region of Soultz, the faulted granitic basement is overlain by 1400 m of sediments. The URG's deep thermal structure, which is likely to be related to mantle uplift, was investigated by deep seismic surveys (Brun and Wenzel, 1991). Relief of the rift borders is still important and the Moho topography shows an important rise up to 24 km depth in the southern URG (Dèzes et al., 2004 and Edel et al., 2007). The shallow heat flow in the graben ranges between 100–120 mW/m^2 (Pribnow and Schellschmidt, 2000). Extensive borehole data show that the temperature within the graben at depths of 1 km is highly variable, the thermal anomaly at Soultz being particularly high. The mineralogical composition of the Soultz granites is relatively poor in radiogenic elements with less than 7 ppm of uranium (Stussi et al., 2002). This is insufficient to explain the geothermal anomaly which is mainly controlled by natural fluid flow. The sediments of the Soultz area were already well explored befthe project began because they host the Pechelbronn-Merkwiller oil field. The Tertiary filling sediments are marine and lacustrine limestones, marls and evaporites, including the petroleum layers of Pechelbronn, unconformably overlying the Jurassic limestones, the German Triassic layers and the Permian sandstones. These Cenozoic, Mesozoic and Permian sediments have been deposited on the Paleozoic basement, which includes porphyritic monzo-granite and two-mica granite. More than 5000 oil wells were drilled prior to 1970, giving an excellent overview of stratigraphy and structures within the post-Paleozoic sediments. Temperature measurements in ~500 of these wells identified the exceptionally high temperature gradients that define the Soultz geothermal anomaly, and showed that the isotherms are primarily influenced by the tectonic structure of the Rhine graben (Haas and Hoffmann, 1929). As can be seen on Fig. 3, there is a strong lateral variation in geothermal gradient, which is especially marked in the vicinity of the fault at Soultz-sous-Forêts where the hottest zone at 400 m depth is located along the western part of the Soultz horst and is characterized by a NE-SW elongation (Gérard et al., 1984). A structural evaluation of this area was performed by (Schnaebele et al., 1948) in 1948 for oil exploitation purposes. They noted that the area was compartmented by normal faults which induce a horst and graben structure.

In the 1980s several oil companies conducted seismic exploration in order to image fault structures within the sedimentary section. Thus, prior to the drilling of GPK-1, the depth of the top of the crystalline basement was known from both seismic reflection profiles and an old oil well drilled close to the site. This well, denoted as 4616, was drilled down to 1403 m and reached the basement at 1380 m. A core taken at the base of this well showed a granite. The first geothermal exploration well at Soultz, GPK-1 (Fig. 1), was drilled into the top of this area in 1987–1988 (Cautru, 1988 and Gérard and Kappelmeyer, 1987). All these data were available and were partly reinterpreted during the preparatory phase of the project, prior to the drilling of GPK-1.

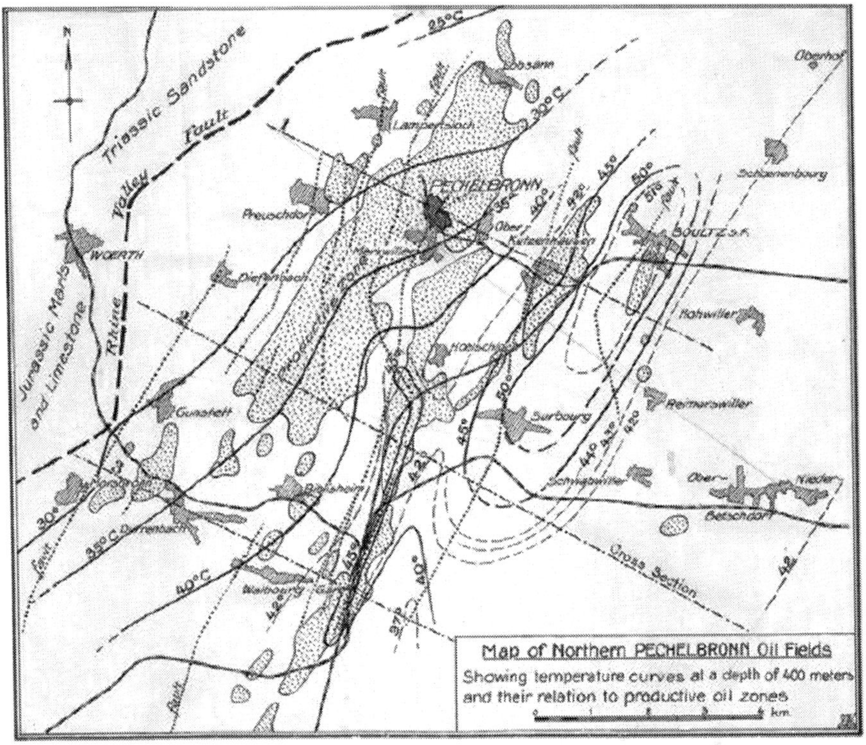

Figure 3: Historical document showing temperature map at 400 m depth based on oil well measurements in the Pechelbronn-Soultz area, from (Haas and Hoffmann, 1929). Hatched areas represent village location and stippled areas represent oil productive zones.

MAIN SCIENTIFIC OUTCOMES FROM EXPLORATION TO RESERVOIR DEVELOPMENT

Geology

The crystalline basement is encountered at about 1.4 km depth below the sedimentary section, and consists of Paleozoic granites that are highly fractured at all scales, from micro-cracks at grain scale to fracture or faulted zones. In the Soultz granite, two main natural fracture organizations are recognized: individual fractures that are seen in cores and borehole image logs that are pervasively distributed in the rock mass, and fracture zones which are larger-scale structures of highly clustered natural fractures that have thicknesses of up to 10–20 m. Fracture zones are observed from various borehole data (core, geophysical logs, borehole image logs) but also imaged far from the wells by vertical seismic profiling (VSP) (Sausse et al., 2010) and their microseismic expression during injections (Dorbath et al., 2009 and Evans et al., 2005b). Hydrothermal alteration is generally intimately associated with fracture zones.

Many small-scale fractures have been identified from coring and imaging logs in the Triassic and Permian sandstones (i.e. Buntsandstein layer) as well as in the granite basement (Genter and Traineau, 1992). They are nearly vertical and have orientations that vary around N175°E in the Buntsandstein sandstones and from N160°E to north-south in the Paleozoic granites. These directions differ from those observed at the surface in this part of the Rhine graben, which are about N20°E to N45°E. In the granites, about 40 fracture zones have been identified between 1400 and 5000 m depth (Dezayes et al., 2010). These are high-angle structures that predominantly strike ~N160°E ± 10°, although secondary sets that strike N20°E ± 10° and N130°E ± 10° are also present. The fracture zones vary in scale from several tens of metres, suggested by core and imaging logging, to at least several hundred meters indicated by microseismic imaging (Evans et al., 2005b, Genter et al., 2000 and Genter et al., 1997). There is also geological and geophysical evidence to suggest that deep wells GPK-2 and GPK-3 may be linked by a fracture zone with an extent of at least 3 km (Sausse et al., 2010). A detailed 3-D analysis of the spatial distribution of fracture zones imaged in the wells has been performed to give a geometrical image of the larger structures in the reservoir (Dezayes et al., 2010 and Sausse et al., 2010). Two petrographically-distinct granite units are encountered in the section to 5 km depth: a grey porphyritic monzo-granite (334 ± 4 Ma) to about 4.5 km depth, and a grey fine-grained two-mica granite (327 ± 7 Ma) below (Cocherie et al., 2004 and Stussi et al., 2002). Some Hydrothermally Altered Fracture Zones (HAFZ) showed a low natural permeability, evidenced by the occurrence of brines in the drilling fluid returns (the formation fluid has 100 g/L of dissolved solids) (Vuataz et al., 1990) and in subsequent hydraulic and geophysical testing (Evans et al., 2005a and Schellschmidt and Schulz, 1991). These HAFZ showed both high fracture density and strong hydrothermal alteration (Genter, 1989). Both fossil and present-day natural fluid circulations in the fractures have resulted in a strong dissolution of the primary minerals such as biotite and plagioclase, and a significant deposition of some altered minerals (Fig. 4) such as clay minerals (illite, tosudite), calcite, secondary quartz and sulfides (Genter and Traineau, 1992). Organic matter was also observed in strongly altered granite in EPS-1 well (Ledésert et al., 1996), indicating hydraulic communication between basement and sediments at some time in the past. Most of these geological results were known after the drilling of GPK-1 down to 2000 m and the coring of EPS-1 to 2230 m. Subsequent drilling campaigns to 3.8 km and eventually to 5000 m depth largely confirmed the previous results, and revealed the two-mica granite below 4.5 km.

Figure 4: Example of hydrothermally altered and fractured granite core taken at 1377 m depth in GPK-1 well showing a nearly vertical sheared fracture.

Stress State

The stress state at Soultz is consistent with a graben setting, with a minimum principal stress that is horizontal and about 53% of the vertical stress. The maximum horizontal principal stress, SH_{max}, is oriented N169°E ± 14°. The large standard deviation is due to variations in stress orientation that occur more or less continuously along the wells reflecting stress perturbations associated with fractures and fracture zones (Valley, 2007). The alignment of the wells at Soultz was chosen to coincide approximately with SH_{max}. The principal stress magnitudes are given as a function of depth z, by (Valley and Evans, 2007):

$$S_v\ [\text{MPa}] = 0.3 + 25.60z\ [\text{km}]$$

$$Sh_{min}\ [\text{MPa}] = -1.8 + 14.09z\ [\text{km}]$$

$$-1.2 + 22.95z\ [\text{km}] \leq SH_{max}\ [\text{MPa}] < -1.4 + 26.775z\ [\text{km}]$$

The vertical stress, S_v is derived by integrating density estimates. The minimum principal horizontal stress, Sh_{min}, is obtained from the maximum pressure attained at the casing shoes of the wells during hydraulic stimulations (Cornet and Bérard, 2003), and conforms well to a linear trend down to the deepest measurement at 4.5 km. The maximum principal stress magnitude, SH_{max}, is constrained by the depth distribution of breakouts and drilling-induced tension fractures (DITFs) to be within 10% of S_v, a result that is consistent with the observation that focal mechanism solutions of micro-earthquakes are a mix of strike-slip and normal (Cuenot et al., 2006b), and with constraints imposed by limits on rock mass strength.

The low value of Sh_{min} implies that shear stresses in the granite are very high. Consequently, many fractures and fracture zones are critically-stressed in as much as they would fail under ambient conditions if their strength was governed by a Coulomb friction criterion with a coefficient of less than 0.95 (Fig. 5 after (Evans, 2005)). The rock mass and the critically-stressed fracture zones within it are clearly very strong, despite the presence of illite within the fracture zones, which is very weak. The strength of the fracture zones is best explained by an internal architecture that involves small-scale fractures connected by relatively intact rock bridges and jogs (Evans, 2005).

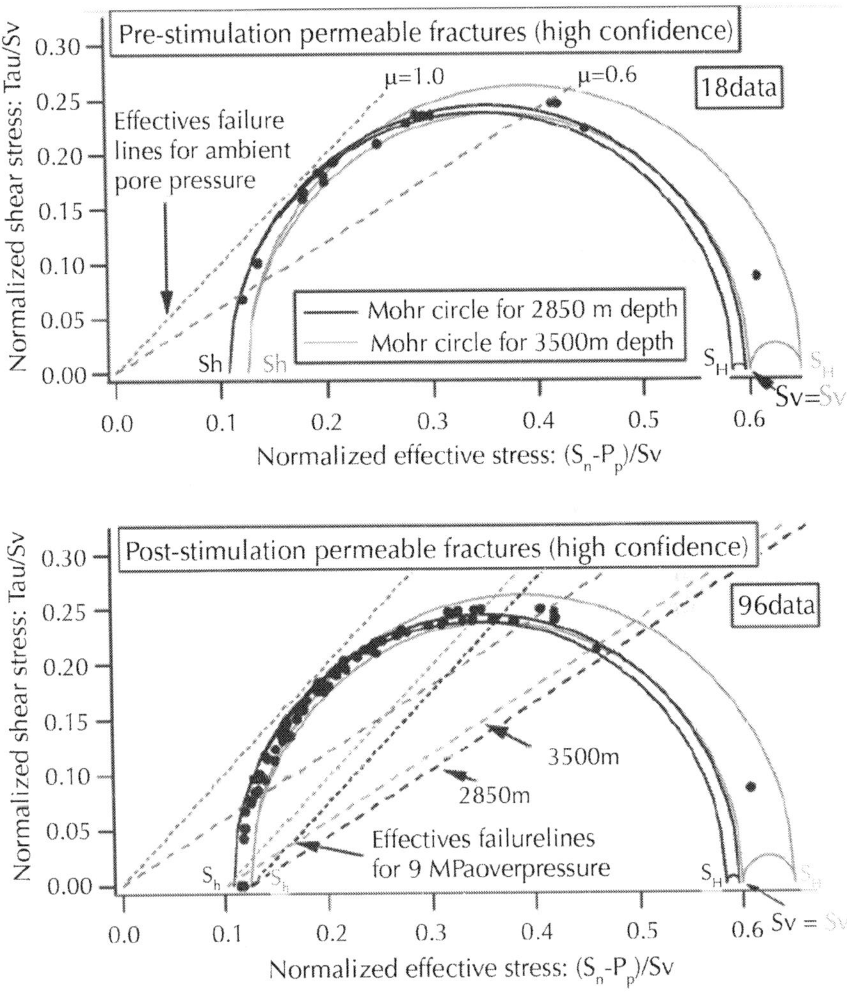

Figure 5: Mohr circles showing the stress state between 2.85 and 3.5 km from (Evans, 2005). (a) shows the shear and normal stress across fractures that were permeable prior to stimulation. Almost all are critically stressed; (b) shows the same as a) but for the post-stimulation permeable fractures. The failure lines for ambient and 9 MPa overpressure (the stimulation pressure) are shown. The vast majority of newly permeable fractures are

critically-stressed. However, many non-permeable fractures (not shown) did not fail but were also critically-stressed.

Fluid Geochemistry

The collection of a sample of native formation fluid that was free of contaminants, such as drilling fluid proved difficult. The most relevant geochemical data on the nature, origin, circulation and deep temperature of these fluids were obtained from GPK-1 and GPK-2 (Aquilina et al., 1997, Pauwels et al., 1993, Sanjuan et al., 2006a and Sanjuan et al., 2010). Samples collected from the GPK-2 well-head and from depths ranging from 650 to 3470 m have similar chemical and isotopic compositions (NaCl brine with a pH value close to 5) and a high salinity (total dissolved solids (TDS) of about 100 g/L), which suggests that fluids at all depths down to 5 km have a common sedimentary origin and have undergone identical fluid-rock interaction processes. Chemical and gas geothermometers, as well as geochemical modelling simulations, suggest that the native geothermal brine and associated gases (predominantly CO_2) are equilibrated with a mineralogical assemblage at temperatures close to 220–240 °C (> 202 °C measured at the GPK-2 bottom-hole in (Baria et al., 2001)). Sodium/lithium geothermometers and isotopic lithium values suggest that these equilibrium reactions occurred in a sedimentary rather than a granite reservoir. These constraints collectively suggest that the brine at Soultz has migrated from a reservoir that is situated more towards the Graben centre, where the Triassic sedimentary formations are the deepest and temperatures of 220–240 °C are expected to occur at 4 km depth (Munck et al., 1979). Such migration could have occurred through a complex system of deep faults that is still poorly constrained. A 3D seismic survey would be very useful for imaging the deepest structures of the Rhine graben in the sediments in order to understand better the geological architecture. A tracer test conducted between 2000 and 2002 in the GPK-2 open-hole section below 4.5 km indicated a natural flow rate of native geothermal brine of 1.0–1.2 m^3/h (Sanjuan et al., 2006a and Sanjuan et al., 2006b), which is identical to that calculated from the fluid flow parallel to the Graben strike, based on a convection model and numerical 3D modelling (Bächler, 2003).

Hydraulics

Hydraulic tests conducted on HAFZs have shown transmissibilities (i.e. the product of permeability and borehole length over which the permeability applies) from zero up to 10^{-11} m^3, whereas the transmissibility of the granite rock mass excluding the high-transmissibility zones is typically 10^{-16} to 10^{-17} m^3 (Evans et al., 2008). The highest transmissibility of 10^{-11} m^3 was measured at a major HAFZ intersected at 2120 m depth in GPK-2 well which, once intersected, resulted in total mud and cuttings loss throughout the subsequent drilling operations. Pre-stimulation testing of the entire GPK-1 open-hole between 2.85 and 3.6 km showed a major transmissive HAFZ at 3.5 km and seven HAFZs of small but non-zero transmissibility spaced every 50–100 m along the well, all of which were critically stressed (Fig. 5) (Evans et al., 2005a). These zones were exclusively activated during the stimulation to become the principal transmissive features in the well. Hydraulic tests performed after hydraulic stimulation showed that the well had been connected to a fault or major fracture zone of large hydraulic capacity through flow paths that emerged into the well at the HAFZs. The hydraulic impedance (i.e. the resistance to flow) of these flow paths exhibited a turbulent-like behaviour that was not limited to the near wellbore region (Evans, 2000 and Kohl et al., 1997), suggesting the flow field was not strongly divergent, probably due to channeling. It is likely that the high-capacity fault to which the well connects via the stimulated flow paths is part of a connected system that allows fluids to move through the rock mass over significant distances. Such features are rarely intersected by vertical boreholes because they are high-angle. These considerations suggest a strategy for establishing a satisfactory hydraulic connection between two boreholes to form a doublet, and that is to link them individually to the connected system of major fracture zones and faults through stimulation operations.

Stimulation

Hydraulic stimulation is the principal reservoir development technique that has been used to increase the low natural permeability of the HAFZs penetrated by the different Soultz wells. The permeability increase is believed to be governed mainly by a shearing mechanism occurring on pre-existing fracture planes, although the details of the geometry of the flow paths produced are uncertain (Evans et al., 2005a, Evans et al., 2005b and Gentier, 2004). For example, shearing of an offset fracture pair can produce a conduit that channels flow normal to the shearing direction (Evans et al., 2005b and Jung and Weidler, 2000). There is also evidence for localised, opening-mode (tensile) fracturing below the casing shoes of some boreholes, where wellbore pressure during stimulation tends to approach the local minimum principal stress level (Cornet and Jones, 1994, Evans et al., 2005a, Jung et al., 2008 and Schindler et al., 2008). However, vertical tensile fractures observed at the borehole walls above 3 km are in most cases induced by cooling during drilling or by swabbing pressures during trips and, whilst good indicators of the directions of the principal horizontal stresses (Valley, 2007), are unlikely to extend significantly from the wellbores.

Almost all wells at Soultz have been subjected to a hydraulic stimulation, and in all but one case the injectivity/productivity index (i.e. the injection flow rate per unit wellhead pressure under steady-state conditions) was significantly improved. For the intermediate depth well GPK-1 stimulated between 2.8 and 3.6 km, the improvement was a factor of 15 (Evans et al., 2005a). For the three deep wells, each stimulated between 4.5 and 5.0 km true vertical depth (TVD), the improvement for GPK-2 and GPK-4 was a factor of about 20, whereas that for GPK-3 was only a factor of ~1.5, although GPK-3 already had the highest initial index prior to stimulation (Nami et al., 2008 and Portier et al., 2009). In all cases, the improvements reflect increased transmissibility of HAFZs, usually at the major altered fractures within the zone. In well GPK-1, these major fractures showed measurable post-stimulation dislocations of millimetres to centimetres (Cornet et al., 1997 and Evans et al., 2005a). Estimation of the post-stimulation injectivity/productivity indices of the three deep wells was complicated by the failure to reach steady-state conditions in the post-stimulation characterization tests. It should be noted that in each of these wells, the downhole injectivity and productivity indices appear to be the same, so the terms are used interchangeably. Thus, single-well injectivity/productivity indices were estimated based upon the wellhead pressure and flow prevailing 3 days only after the start of the test, with the other wells shut-in (Nami et al., 2008). The measured indices were 4.0 l/s/MPa for GPK-2, 3.5 l/s/MPa for GPK-3, and 2.0 l/s/MPa for GPK-4.

A variety of chemical stimulations were conducted on all deep wells after the hydraulic stimulations (Nami et al., 2008 and Portier et al., 2009). The improvement in 3-day injectivity/productivity index by the acid treatments was a factor of about 1.25 for GPK-2, about 1.15 for GPK-3, and about 2.5 for GPK-4 (Table 2). It should be noted, however, that the large increase in productivity for GPK-4 was largely due to the development of leaks in the casing whose origin is currently uncertain. If only the open hole is considered, the improvement achieved by the acid treatments was a factor of 1.5–1.75. The net 3-day single-well injectivity/productivity indices after the acidizations were: 5.0 l/s/MPa for GPK-2, 4.0 l/s/MPa for GPK-3, and 5.0 l/s/MPa for GPK-4.

Table 2: Summary of the chemical stimulations performed in the three 5-km deep Soultz from (Portier et al., 2009)

Well	**Date**	**Concentration of chemical agents**	**Stimulation results**
GPK2	February 2003	One test in two steps: HCl 0.09% and HCl 0.18%	Well-head pressure drop and productivity increase $0.5L.s^{-1}bar^{1}$
GPK3	June 2003	HCl 0.45%	Injectivity $0.35l.s^{-1}bar^{1}$
	February 2007	OCA (organic clay acid) HT	Weak impact $0.4L.s^{-1}bar^{1}$

GPK4	February 2005	HCl 0.2%	Productivity: 0.2–0.3L. $s^{-1}bar^{-1}$
	May 2006	Preflush: HCl 15% RMA (regular mud acid) HCl 12%-HF 3%	Maximum enhancement of injectivity: 35%
	October 2006	NTA (nitrilotriacetic acid) 19%	The formation of a plug increased wellhead pressure - Productivity: 0.3–0.4L. $s^{-1}bar^{-1}$ (after RMA and NTA treatments)
	March 2007	OCA HT	Productivity: 0.4–0.5L. $s^{-1}bar^{-1}$

Microseismicity

Microseismic activity was monitored with both down-hole and surface networks during all stimulations to map the locations where shear failure was occurring (Cuenot et al., 2006a and Horalek et al., 2008). Typically, several thousand microseismic events would be recorded by the down-hole seismic network during a stimulation, and somewhat fewer by the surface seismic network (Baria et al., 2004, Cuenot et al., 2008a, Charléty et al., 2007 and Dorbath et al., 2009). High-precision mapping of the seismicity associated with the injection into GPK-1 illuminated the large scale geometry of HAFZs within the active volume and provided insights into their internal microstructure (Evans et al., 2005b). Comparison of the seismic responses to stimulation of the three deep wells showed that they are dependent to some degree on the nature of the HAFZs encountered in each of the wells (Dorbath et al., 2009). The open-hole section of deep well GPK-2 is characterized by a dense network of medium-scale fracture zones, and the seismic cloud is rather compact and includes ~700 events of magnitude higher than 1 (Fig. 6). The initial injectivity/productivity index of this well was low, but was increased by a factor of 20 by the stimulation. The open hole section of well GPK-3 is hydraulically dominated by a very large HAFZ at 4.7 km measured depth (MD) that absorbed about 70% of the injected fluid and gave the well a high initial injectivity of 3 l/s/MPa. Stimulation produced 250 events larger than magnitude 1.0 (Fig. 7). However, the proportion of events with magnitudes larger than 2 was greater for GPK-3 than for GPK-2. This observation could indicate that the characteristics of the seismicity around GPK-3 are influenced by the large-scale fracture zone. However, the increase in injectivity/productivity index after stimulation was small.

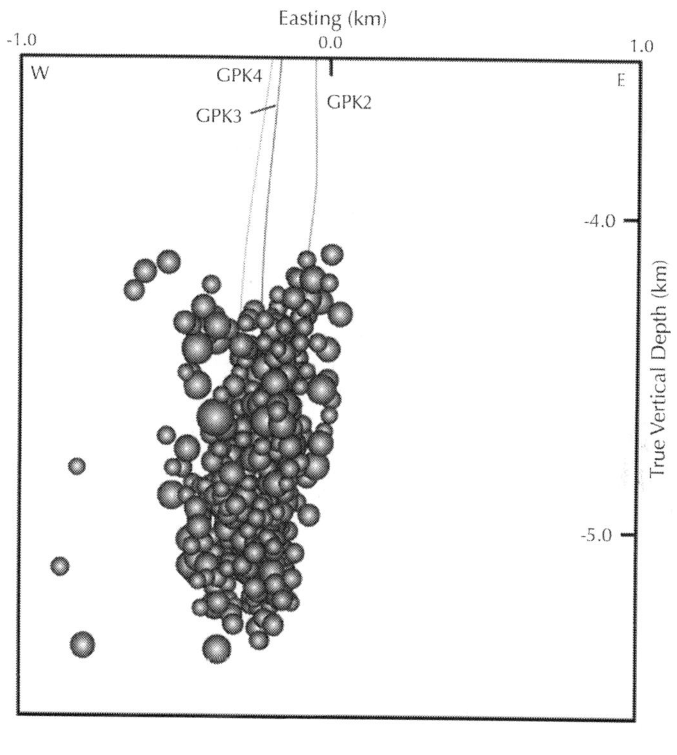

Figure 6: The 2000 stimulation of GPK-2: Vertical cross-section through the cloud of induced seismic events ($M > 1$) along a line N20°W from (Dorbath et al., 2009). The magnitude scale for the microseismic events is plotted on Fig. 7.

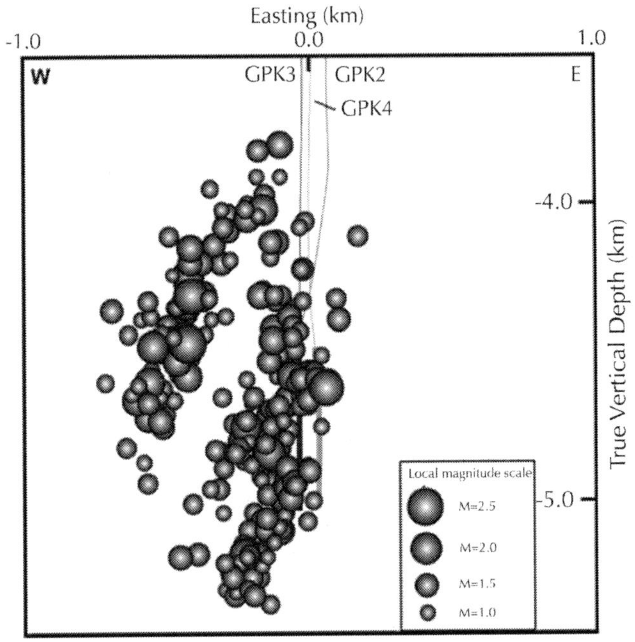

Figure 7: The 2003 stimulation of GPK-3. Vertical cross section through the cloud of induced seismic events ($M > 1$) along a line N9°W (Dorbath et al., 2009).

Several seismic events induced during stimulation operations attained a magnitude 2.0, the maximum being 2.9 during the injection into GPK-3, and were felt by the neighbouring population, provoking unwanted alarm. The highest magnitude events tended to occur shortly after the end of injection, during the shut-in phase. Studies to better understand injection-induced microseismicity, particularly with regard to the generation of the larger events, are on-going. It is considered very important to identify ways of avoiding disturbance to populations living close to geothermal projects that involve hydraulic stimulation operations.

Thermics

Temperature measurements were routinely performed in all wells at Soultz. The temperature profile to 5 km depth measured in well

GPK-2 at conditions close to thermal equilibrium is presented in Fig. 8. The profile is typical of all deep wells, with temperatures of 200 °C being reached at 5 km depth. The localized disturbances at 2 km and 3.4 km reflect remnant cooling from drilling or from stimulation injections conducted at those depths. The profile can be divided into three sections with different geothermal gradients reflecting different heat transport processes. The upper section lies entirely in the sediments from the surface to 1 km depth and has a very high gradient of about 110 °C/km, indicating heat transport is predominantly conductive. The intermediate depth section from 1 km to about 3.3 km depth is characterized by a very low gradient of about 5 °C/km, which suggests heat transport is dominated by advection, probably in the up-flowing limb of a convective system that is active in the granite and Triassic sandstones (Le Carlier et al., 1994). This implies significant natural fluid movement is occurring within faults and fracture zones, at least between 1.0 and 3.3 km (Fig. 8). Below 3.3 km, the geothermal gradient increases again to 30 °C/km and becomes linear with depth, indicating a return to a conduction-dominated heat flow regime. This might be explained as indicating that faults and fracture zones have insufficient connected permeability below 3.3 km depth to allow convection to develop (Le Carlier et al., 1994 and Pribnow and Schellschmidt, 2000).

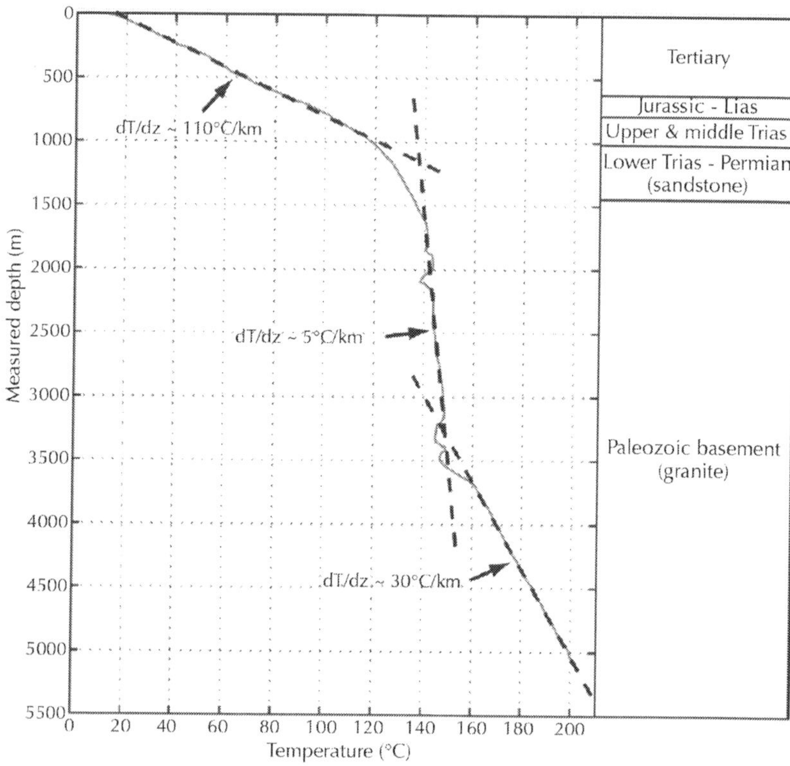

Figure 8: Equilibrium temperature profile obtained from log run in GPK-2 several months after drilling operation finished. (Data from LIAG-Hannover, Germany). Main geological units are presented versus depth.

PERFORMANCE OF THE UPPER AND LOWER SYSTEMS DURING CIRCULATION

The 1991–1998 phase of the Soultz project culminated in the 4-month continuous circulation of the upper-reservoir doublet system formed by the open hole sections of GPK-1 and GPK-2 which were 450 m apart (Aquilina et al., 1998, Baria et al., 2001,Baumgaertner et al., 1998 and Bruel, 2002). Several limited-duration circulation tests have also been performed in the lower reservoir to date: without downhole pumps in 2005, and with one downhole pump and with power generation in 2008 (Cuenot et al., 2006a, Gérard et al., 2006, Sanjuan et al., 2006a and Sanjuan et al., 2006b). The main results of the circulation tests are described in this section.

The upper reservoir between 3.0 and 3.6 km depth was the first truly engineered geothermal system to be constructed at Soultz (Fig. 1). In 1997, this reservoir was successfully circulated for four months at a mass flow rate of 25 kg/s in closed-loop mode at fluid pressures far below the stimulation pressure levels. A down-hole pump was installed at 430 m depth in the production well (GPK-2) to improve recovery by taking advantage of the natural far field permeability. Fluid mass balance (i.e. zero fluid loss) was maintained throughout the circulation with minimal energy consumption. During this experiment, 244,000 tons of brine at temperatures up to 142 °C were produced from GPK-2 and re-injected in GPK-1 at a temperature of 40 °C using a centrifugal pump. The GPK-1 wellhead injection pressure declined from 4 MPa to 2 MPa during the test (Aquilina et al., 1998 and Baumgaertner et al., 1998). The decline was slow and steady for the first half of the test, but then underwent a rapid drop after 60 days, an event that coincided with the termination of scaling inhibitor injection. Flow logs indicate that the gradual change reflected declining impedance of a HAFZ at 3225 m (Evans et al., 1998), whereas the steep drop was largely due to an impedance drop at a HAFZ just below the casing shoe. Both have been ascribed to rock shrinkage due to cooling (Evans et al., 1998 and Bruel, 2002). By the end of the circulation, the system circulation impedance was only 0.1 MPa/l/s. This was the first time that this long-standing commercial target had been attained in any EGS system worldwide, and then for the largest well separation of 450 m ever attempted (Evans et al., 2008). The specific gravity of the production fluid increased during the test from 1.048 g/cm^3 in July to 1.063 g/cm^3 in November 1997 (Rummel et al., 1997), reflecting a progressively larger fraction of formation water in the produced fluid. This, together with the relatively small quantities of tracer (i.e. 30%) recovered from tests, indicates that production was accessing fluid stored in the 'far-field', probably in the network of major fracture zones and faults present in the rock mass (Fig. 9). Nevertheless, the tracer results also showed a rapid response, indicating that at least some of the principal flow paths linking the wells were direct. Application of a discrete fracture network model to the multi-scale fractured reservoir conditioned by the tracer response suggested that a 5% production temperature decrease could take place after 20 years of operating the doublet at a circulation rate of 25 l/s (Bruel, 2002). The usable thermal power produced for a re-injection temperature of 40 °C attained ~11 MW thermal by the end of the 1997 test.

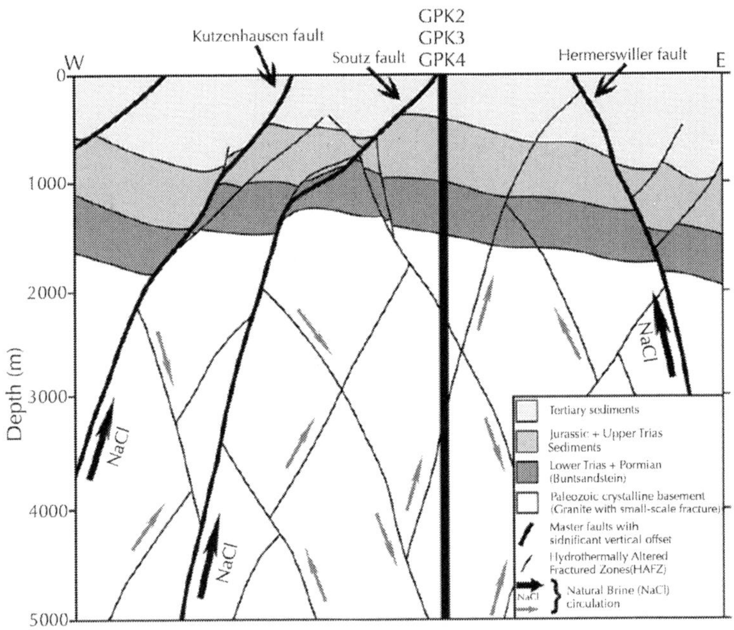

Figure 9: Conceptual schematic West-East geological cross-section of the Soultz horst based on (Cautru, 1988), showing the main geological units and a conceptual view of the deep, interconnected faults and fractures system. The thick, black, vertical line represents a simplified projection of the geothermal well trajectories. Natural NaCl-rich brine circulations are indicated by black and grey arrows.

The first circulation test of the triplet of wells penetrating the lower reservoir (4.5–5.0 km) (Fig. 1, Table 1) took place for 5 months between July and December 2005. The wells had been hydraulically and chemically stimulated by that time, although some further acidization treatments of GPK-3 and GPK-4 were performed subsequently. A total of 209,000 m^3 of hot brine was produced by buoyancy drive at a cumulative flow rate of 15 l/s from the lateral boreholes GPK-2 (160 °C) and GPK-4 (120 °C) without any downhole production pumps (Gérard et al., 2006). The produced fluid was re-injected into the central well GPK-3 at a wellhead pressure which increased from 4 to 7 MPa. Tracer tests conducted during the circulation showed that ~25% of the injected tracer was recovered from GPK-2 but only 2% from GPK-4 (Sanjuan et al., 2006a and Sanjuan et al., 2006b). This asymmetrical response reflects the complex organisation of fracture zones or faults describing different underground fluid circulation loops, the hydraulic connections between GPK-3 and GPK-2 being much more direct and faster than those between GPK-3 and GPK-4 (Sanjuan et al., 2006a and Sanjuan et al., 2006b). During this circulation, and all production tests conducted in the lower reservoir, tracer tests and geochemical data invariably showed the presence of the native geothermal brine in the discharged fluids, even after large amounts of external water had been injected into the wells (Sanjuan et al., 2006a and Sanjuan et al., 2006b). This again points to the conclusion that the exchanger is connected to a deep natural reservoir. Some 600 microseismic events were recorded in the 6 months during and immediately following the circulation. Several exceeded magnitude 2.0 and some of these were felt (Cuenot et al., 2008b).

In 2008, the lower reservoir was again circulated during the power-plant commissioning phase. This time, a line-shaft production pump was operational at 350 m depth in GPK-2, with GPK-4 remaining shut-in (Table 1). Since the earlier circulation in 2005, injection well GPK-3 had been subject to a further acid stimulation in 2007, which had little effect on injectivity (Nami et al., 2008), whilst well GPK-4 had been subject to three different types of acidization treatments with considerable success (Table 2). Circulation began at the end of June 2008 and lasted until mid-August 2008. During this period, the pump-assisted production from GPK-2 was around 25 l/s at a temperature of 155 °C. The production fluid at the surface was maintained at a pressure of 2 MPa before passing through a pump for injection into GPK-3. Wellhead injection pressure began at 6 MPa and increased continuously albeit progressively more slowly to stabilize at 7 MPa for last week of the test. Approximately 190 microearthquakes were associated with the circulation, which gives an event rate comparable to that observed in the 2005 circulation (Cuenot et al., 2008b). They also occurred in much the same locations as the earlier events, but the magnitude did not exceed 1.4, in contrast to the earlier events, several of which exceeded 2.0 (Cuenot et al., 2006b).

The lower magnitudes may reflect several differences between the two tests: the duration in 2005 was 6 months as compared to 2 months in 2008; a larger volume of water was circulated in 2005; and the use of a down-hole production pump in 2008 induced higher production flow rates and lower net reservoir pressure compared to the artesian production from two production wells in 2005. Additionally, between 2005 and 2008, the wells were chemically stimulated (Nami et al., 2008). Moreover most of the larger magnitude earthquakes in 2005 occurred after a sudden increase of injection flow rate and pressure. This behaviour is quite similar to the tendency of large events to occur during the shut in phase of stimulation tests (Cuenot et al., 2006a), although the underlying mechanisms are not necessarily identical.

THE SOULTZ RESERVOIR: A CHANNEL NETWORK POORLY CONNECTED THROUGH HYDROTHERMALLY ALTERED AND FRACTURED ZONES

By drilling into the Soultz granite to 5 km depth, several depth sections containing permeable fractures were observed (Dezayes et al., 2010). This vertical zonation is organized in three main sections from the top of the basement to TD. The main HAFZs are located: (1) in a highly fractured and altered granite interval between 1.4 and 2.2 km (GPK-1, 1820 m; EPS-1, 2180 m; GPK-2, 2120 m); (2) in the upper reservoir between 3.5 and 3.9 km depth (GPK-1, 3490 m; GPK-2, 3900 m); and (3) in the lower reservoir below 4.5 km depth (GPK-3, 4770 m). This vertical zonation corresponds to the geometrical intersection of the drillholes with major nearly vertical fracture zones. As all the wells are roughly aligned along the N170°E striking direction and as the major HAFZs range azimuthally between N150°E and N170°E, they appear in cross-section in 3 main depth sections. The presence of additional fracture zones which do not intersect the wells but are near them is indicated from microseismic imaging (Evans et al., 2005b, Moriya et al., 2002 and Phillips, 2000) and Vertical Seismic Profiling (VSP) studies (Place et al., 2007). Moreover, there are several major fracture zones that are not strongly hydraulically active. For example, the fracture zone in GPK-1 at 3.5 km, whilst transmissive, is not the principal hydraulically-active HAFZ in the well after stimulation, even though it was the principal hydro-geologic structure in the open hole section prior to stimulation (Evans et al., 2005b).

The drilling of numerous wells at Soultz has provided a good description of structures present within the rock mass to 5 km depth. As noted earlier, these can be broadly viewed as falling into two categories: small-scale fractures (meter-scale) that are widely distributed in the massive granite but are not hydraulically significant; and more localized, large-scale (i.e. > 100 m) hydrothermally-altered fracture zone (HAFZ) structures created or reactivated by the Rhine graben tectonics. Both networks are steeply dipping and show a strong orientation anisotropy with similar strikes of N170°E ± 30°. Numerous HAFZs are intersected by the wells, and these are often found to be naturally permeable, although usually only slightly so. However, several of the HAFZs are clearly major structures in both a geological and hydrological sense. Structures recognised as major in both senses are listed in Table 1 of reference (Dezayes et al., 2010). These major structures will be expected to have significant influence on the pattern of forced fluid flow within reservoirs developed in their vicinity, although they may not be dominant since other sub-vertical major HAFZ structures may be present that are not intersected by any well. High-resolution microseismic imaging of the microseismic cloud generated during the stimulation of the upper reservoir demonstrates that many conductive features that do not intersect the wellbore are activated during stimulation (Moriya et al., 2002 and Phillips, 2000). Most likely, almost all primary flow is confined to HAFZs, both before, during and after stimulation (Evans et al., 2005b). Within a given HAFZ, the natural permeability structure is probably channelled with a rather complex 3D organization (Caine et al., 1996).

Exploration and testing of the Soultz granite has shown that it differs significantly from the original concept of a HDR reservoir (Smith, 1983). The presence of the HAFZs, which together form a connected system of large hydraulic capacity (i.e. storage), means that the Soultz reservoirs share many facets of conventional geothermal reservoirs that benefit from re-injection. It is probable that the elements of the natural reservoirs govern the flow field that develops under forced flow conditions within the reservoirs.

THE SOULTZ POWER PLANT

Based on the results of the circulation test performed in 2005, and the subsequent improvement of the hydraulic performance of the three deep wells, it was decided to build a 1.5 MWe geothermal power plant. In view of the highly saline geothermal fluid which prohibits direct vaporization in a turbine, and the expected production temperature, it was decided to use an Organic Rankine Cycle (ORC) system that used a low boiling-point working fluid (isobutane) in the heat-power conversion scheme. The turbine is radial and operates at around 13000 rpm. The generator is asynchronous, operates at around 1500 rpm, and has an 11 kV output, which is transformed up to feed into the 20 kV local power network. The geothermal fluid is cooled down to 80–90 °C in the heat exchangers of the binary unit before reinjection. The system is constructed to allow the production from either or both of the two wells to be used to feed the power production loop. The expected production flow rate from both wells together is about 35 l/s for a production temperature of 175 °C. As of September 2009, the power plant has been assembled and produced electrical power during the circulation in June–August 2008 when production came from GPK-2 with a Line-Shaft Pump (LSP). Since then, an Electro-Submersible Pump (ESP) has been deployed in the second production well GPK-4. The next phase of the project will involve the long-term circulation of the system with both production wells operational, in order to optimize the power plant operation and to progressively increase the generated electric power, which will be fed into the local network. This phase will begin shortly.

FUTURE GEOTHERMAL TARGETS IN THE RHINE GRABEN

Exploration at the Soultz site has shown that the geothermal anomaly and the presence of oil in the Pechelbronn oil field are related, and that the heat distribution is mainly controlled by large-scale convection of high-salinity fluid occurring within fractured formations (sandstones, granite). The granite near the top of the basement (1.4 to 2.2 km depth) is a promising geothermal target because it is more naturally fractured and more permeable than the upper and the lower reservoirs. The nature of the deeper fractured basement at Soultz is well documented along the boreholes, but the inter-well domain is poorly constrained because all holes are near-vertical or steeply inclined. This borehole geometry was driven by the fact that the orientation of the main fracture system was aligned with the maximum horizontal principal stress (Baumgaertner et al., 1998). Since it is difficult to image the fracture system in 3D prior to drilling, it might be more convenient for future geothermal projects in a similar geology/stress context to drill inclined holes perpendicular to the strike of the fracture system to maximise the likelihood of intersecting as many as possible. Such an approach has recently been successfully applied in a Soultz spin-off project near Landau, Rhineland Palatinate, Germany, some 40 km north of Soultz, where two shallow deviated wells were drilled into the lowermost sediments and the top of the basement (Baumgaertner et al., 2007).

Based on geological findings and drilling results, the Soultz adventure contributed significantly to improving our vision of the EGS potential and will contribute to the design of new projects in France and abroad (Ledru, 2008).

CONCLUSIONS

Some 20 years of EGS research at the Soultz geothermal site has produced numerous scientific results and advances related to drilling, surface exploration methods, basement geology, geochemical and geophysical monitoring, tracer testing, hydraulic and chemical stimulations, and general EGS know-how. A key finding of the drilling and testing to 5 km depth is that the rock mass contains numerous hydrothermally-altered fracture zones that host large volumes of brines and essentially constitute a significant natural hydraulic reservoir. These high-angle zones, which strike principally to the NNW, are seen on core and image logs and can be mapped as discrete structures at scales of up to several hundred meters on high-resolution images of the pattern of microseismicity and VSP surveys. The structures control the movement of fluids through the rock mass under natural and forced fluid flow conditions, and thus determine the behaviour of the reservoir. Flow between the wells during circulation of the two systems built at 3.5 km and 5 km depth in large part followed natural flow paths defined within these structures, leading to mixing of the circulation fluid with reservoir brines. The hydraulic and chemical stimulations served to enhance the transmissivity of these natural flow paths. The 3.5 km system was the first EGS to have attained a circulation impedance of 0.1 MPa/l/s. The long-term circulation of the 5 km system is about to begin. The future development of EGS projects in Soultz-like environments would benefit from improved exploration methods for imaging structures remote from the wells. It is also important to obtain a better understanding of the factors that promote stimulation/circulation-related microseisms large enough to be felt by local population, and to identify ways to mitigate the hazard.

ACKNOWLEDGMENTS

This work was performed within the framework of the European EGS Pilot Plant project which was supported mainly by the European Commission, BMU (Germany), ADEME (France), the Swiss Federal Offices of Education and Research (SER), and Energy (SFOE), and by a consortium of industrial members (EDF, EnBW, ES, Pfalzwerke, Evonik). The manuscript greatly benefited from the constructive remarks of John Garnish and Patrick Ledru.

REFERENCES

1. Aquilina, L., Pauwels, H., Genter, A., Fouillac, C., 1997. Water-rock interaction processes in the Triassic sandstone and the granitic basement of the Rhine Graben: geochemical investigation of a geothermal reservoir. Geochim. Cosmochim. Acta 60, 4281–4295.
2. Aquilina, L., Rose, P., Vaute, L., Brach, M., Gentier, S.,

Jacquot, E., Audigane, P., Jeannot, R., Tran-Viet, T., Jung, R., Baumgaertner, J., Baria, R., Ge´ rard, A., 1998. A tracer test at the Soultz-sous- Fore^ts Hot Dry Rock geothermal site. In: 23rd Workshop on Geothermal Reservoir Engineering, Stanford University, California, USA, pp. 343–350.

3. Ba¨chler, D., 2003. Coupled thermal-hydraulic-chemical modeling at the Soultz-sous-Fore^ts HDR reservoir (France), Thesis Doctor of Natural Sciences, Swiss Federal Institute of Technology, Zu¨ rich, 151 p.
4. Baria, R., Baumgaertner, J., Ge´ rard, A., Jung, R., Garnish, J., 1999. European HDR research programme at Soultz-sous-Fore^ts (France) 1987–1996. Geothermics 28, 655–669.
5. Baria, R., Baumgaertner, J., Ge´ rard, A., Weidler, R., Hopkirk, R. et al., 2001. European HDR geothermal research programme, 1998–2001, EU final report, Contract no (JOR3-CT98-0313).
6. Baria, R., Michelet, S., Baumgaertner, J., Dyer, B., Ge´ rard, A., Nicholls, J., Hettkamp, T., Teza, D., Soma, N., Asanuma, H., Garnish, J., Me´gel, Th., 2004. Microseismic monitoring of the world's largest potential HDR reservoir. In: 29th Workshop on Geothermal Reservoir Engineering, Stanford University, California, USA, pp. 194–201.
7. Baumgaertner, J., Ge´ rard, A., Baria, R., Jung, R., Tran-Viet, T., Gandy, T., Aquilina, L., Garnish, J., 1998. Circulating the HDR reservoir at Soultz: maintaining production and injection flow in complete balance: initial results of 1997 experiment. In: 23rd Workshop on Geothermal Reservoir Engineering, Stanford University, California, USA, pp. 11– 20.
8. Baumgaertner, J., Menzel, H., Hauffe P., 2007. The geox GmbH Project in Landau - The first geothermal power plant project in Palatinate / Upper Rhine Valley, First European Geothermal Review, Geothermal Energy for Power Production, October 29 – 31, Mainz, Germany, 33.
9. Bourgeois, O., Ford, M., Diraison, M., Le Carlier de Veslud, C., Gerbault, M., Pik, R., Ruby, N., Bonnet, S., 2007. Separation

of rifting and lithospheric folding signatures in the NW-Alpine foreland. Int. J. Earth Sci. 96, 1003–1031.

10. Bruel, D., 2002. Impact of induced thermal stress during circulation tests in an engineered fractured geothermal reservoir. Oil Gas Sci. Technol. Rev. IFP 57, 459–470.
11. Brun, J.-P., Wenzel, F., the ECORS-DEKORP team, 1991. Crustal scale structure of the southern Rhine graben from ECORS-DEKORP seismic reflection data. Geology 19, 758–762.
12. Caine, J.S., Evans, J.P., Forster, C.B., 1996. Fault zone architecture and permeability structure. Geology 24, 1025–1028.
13. Cautru, J.P., 1988. Coupe ge´ologique passant par le forage GPK-1 cale´e sur la sismique re´flexion, document BRGM. Institut Mixte de Recherches Ge´othermiques, Rapport interne.
14. Charle´ty, J., Cuenot, N., Dorbath, L., Dorbath, C., Haessler, H., Frogneux, M., 2007. Large earthquakes during hydraulic stimulations at the geothermal site of Soultz-sous-Fore^ts. Int. J. Rock Mech. Min. Sci. 44, 1091–1105.
15. Cocherie, A., Guerrot, C., Fanning, C.M., Genter, A., 2004. Datation U-Pb des deux facie` s du granite de Soultz (Fosse´ Rhe´nan, France). C. R. Geoscience 336, 775–787.
16. Cornet, F.H., Be´ rard, T., 2003. A case example of integrated stress profile evaluation. In: 3rd International Symposium on Rock Stress, Balkema, Kumamoto, Japan, 2–4 November.
17. Cornet, F.H., Helm, J., Poitrenaud, H., Etchecopar, A., 1997. Seismic and aseismic slip induced by large fluid injections. Pure Appl. Geophys. 150, 563–583.
18. Cornet, F.H., Jones, R.H., 1994. Field evidence on the orientation of forced water flow with respect to the regional principal stress directions. In: 1st North American Rock Mechanics Symposium, Balkema, Austin, Texas, USA.
19. Cuenot, N., Charle´ty, J., Dorbath, L., Dorbath, C., Ge´ rard, A., 2006a. 2005 circulation experiments: views of

the microseismic activity under production conditions. In: EHDRA Scientific Conference. Soultz-sousFore^ts, France, June 6 p.

20. Cuenot, N., Charle´ty, J., Dorbath, L., Haessler, H., 2006b. Faulting mechanisms and stress regime at the European HDR site of Soultz-sousFore^ts, France. Geothermics 35, 561–575.
21. Cuenot, N., Dorbath, C., Dorbath, L., 2008a. Analysis of the microseismicity induced by fluid injections at the Hot Dry Rock site of Soultz-sousFore^ts (Alsace, France): Implications for the characterization of the geothermal reservoir properties. Pure Appl. Geophys. 165, 797–828.
22. Cuenot, N., Dorbath, L., Frogneux, M., Jund, H., 2008b. Microseismic activity during circulation in summer 2008 and comparison with 2005 circulation test. In: EHDRA Scientific Conference, Soultz-sousFore^ts. France, September 6 p.
23. Dezayes, C., Genter, A., Valley, B., 2010. Structure of the low permeable naturally fractured geothermal reservoir at Soultz. C. R. Geoscience 342, 517–530.
24. De`zes, P., Schmid, S.M., Ziegler, P.A., 2004. Evolution of the European Cenozoic Rift System: interaction of the Alpine and Pyrenean orogens with their foreland lithosphere. Tectonophysics 389, 1–33.
25. Dorbath, L., Cuenot, N., Genter, A., Frogneux, M., 2009. Seismic response of the fractured and faulted granite to massive water injection at 5 km depth at Soultz-sous-Fore^ts (France). Geophys. J. Int. 177, 653–675.
26. Edel, J.-B., Schulmann, K., Rotstein, Y., 2007. The Variscan tectonic inheritance of the Upper Rhine Graben: evidence of reactivations in the Lias, Late Eocene-Oligocene up to the Recent. Int. J. Earth Sci. 96, 305–325.
27. Evans, K.F., 2000. The effect of the 1993 stimulations of well GPK-1 at Soultz on the surrounding rock mass: evidence for the existence of a connected network of permeable fractures. In: World Geothermal Congress 2000, Kyushu – Tohoku, Japan, May 28–June 10, pp. 3695– 3700.

28. Evans, K.F., 2005. Permeability creation and damage due to massive fluid injections into granite at 3.5 km at Soultz: Part 2 - Critical stress and fracture strength. J. Geophys. Res. 110, 14.

29. Evans, K.F., Genter, A., Sausse, J., 2005a. Permeability creation and damage due to massive fluid injections into granite at 3.5 km at Soultz: Part 1 - Borehole observations. J. Geophys. Res. 110, 19.

30. Evans, K.F., Hopkirk, R., Jung, R., Nami, P., Schindler, M., Teza, D., Tischner, T., 2008. Milestone events and key observations in Thermics, Stress and Hydraulics at Soultz (1987–2002). In: EHDRA Scientific Conference, Soultz-sous-Fore^ts. France, September 6 p.

31. Evans, K.F.,

32. Kohl, T., Rybach, L., 1998. Analysis of the hydraulic behaviour of the 3.5 km deep reservoir during the 1995–1997 test series, and other contributions to the European Hot Dry Rock Project, Soultzsous-Fore^ts, France, Report to Swiss Office of Education and Research under contract BBW 95.0673-2, Swiss Federal Institute of Technology.

33. Evans, K.F., Moriya, H., Niitsuma, H., Jones, R.H., Phillips, W.S., Genter, A., Sausse, J., Jung, R., Baria, R., 2005b. Microseismicity and permeability enhancement of hydrogeologic structures during massive fluid injections into granite at 3 km depth at the Soultz HDR site. Geophys. J. Int. 160, 388–412.

34. Genter, A., 1989. Ge´othermie Roches Chaudes Se` ches: le granite de Soultz-sous-Fore^ts (Bas Rhin, France). Fracturation naturelle, alte´ rations hydrothermales et interaction eau – roche, PhD thesis, Universite´ d'Orle´ans, France, 201 p.

35. Genter, A., Traineau, H., 1992. Borehole EPS-1, Alsace, France; preliminary geological results from granite core analyses for hot dry rock research. Scient. Drill. 3, 205–214.

36. Genter, A., Castaing, C., Dezayes, C., Tenzer, H., Traineau, H., Villemin, Th., 1997. Comparative analysis of direct (core) and

indirect (borehole imaging tools) collection of fracture data in the Hot Dry Rock Soultz reservoir (France). J. Geophys. Res. 102, 15419–15431.

37. Genter, A., Traineau, H., Bourgine, B., Lede´ sert, B., Gentier, S., 2000. Over 10 years of geological investigations within the European Soultz HDR project, France. In: World Geothermal Congress 2000, KyushuTohoku, Japan, May 28–June 10, pp. 3707–3712.

38. Gentier, S., 2004. Les massifs rocheux fracture´ s : de la fracture a` la ge´othermie, Me´moire d'habilitation a` diriger des recherches (HDR). Universite´ Bordeaux 1, France.

39. Ge´ rard, A., Kappelmeyer, O., 1987. The Soultz-sous-Fore^ts project: Proceedings of the first EEC/US workshop on geothermal Hot Dry Rock technology. Geothermics 16, 393–399.

40. Ge´rard, A., Menjoz, A., Schwoerer, P., 1984. L'anomalie thermique de Soultz-sous-Fore^ts. Geotherm. Actual. 3, 35–42.

41. Ge´rard, A., Genter, A., Kohl, T., Lutz, P., Rose, P., Rummel, F., 2006. The deep EGS (Enhanced Geothermal System) project at Soultz-sous-Fore^ts (Alsace, France). Geothermics 35, 473–483.

42. Haas, J.-O., Hoffmann, C.R., 1929. Temperature gradient in Pechelbronn oil bearing region, lower Alsace: its determination and relation to oil reserves. Am. Ass. Petrol. Geol. Bull XIII, 1257–1273.

43. Horalek, J., Jechumtalova, Z., Dorbath, L., Sileny, J., 2008. Shear versus nonshear components in source mechanisms of micro-earthquakes induced in hydraulic fracturing experiment at the HDR site SoultzA. Genter et al. / C. R. Geoscience 342 (2010) 502–516 515 sous-Fore^ts (Alsace) in 2003. In: EHDRA Scientific Conference, Soultzsous- Fore^ts. France, September 8 p.

44. Jung, R., Tischner, T., Baumgaertner, J., Teza, D., Nami, P., Schindler, M., 2008. Review of hydraulic stimulation tests

and hydraulic experiments at Soultz. In: EHDRA Scientific Conference. Soultz-sous-Fore^ts, France, September 2 p.

45. Jung, R., Weidler, R., 2000. A conceptual model for the stimulation process of the HDR-system at Soultz. In: Geothermal Resources Council Transactions, Geothermal Resources Council, San Francisco, California, USA, 24–27 September, vol. 24, pp. 143–147. Kohl, T., Evans, K.F., Hopkirk, R.J.,

46. Jung, R., Rybach, L., 1997. Observation and simulation of non-Darcian flow transients in fractured rock. Water Resour. Res. 33, 407–418.

47. Le Carlier, C., Royer, J.-J., Flores, E.L., 1994. Convective heat transfer at the Soultz-sous-Fore^ts geothermal site: implication for oil potential. First Break 12, 553–560.

48. Lede´ sert, B., Joffre, J., Amble` s, A., Sardini, P., Genter, A., Meunier, A., 1996. Organic matter in the Soultz HDR granitic thermal exchanger (France): natural tracer of fluid circulations between the basement and its sedimentary cover. J. Volcanol. Geotherm. Res. 70, 235–253.

49. Ledru, P., 2008. The final conference of the ENGINE Coordination Action: a milestone towards EGS demonstration projects, From. In: Calcagno, P., Slaupia, S. (Eds.), Actes/ Proceedings of the ENGINE Final Conference, 12–15 February, Vilnius, Lithuania, pp. 11–12.

50. Moriya, H., Nakazato, K., Niitsuma, H., Baria, R., 2002. Detailed fracture system of the Soultz-sous-Fore^ts HDR field evaluated using microseismic multiplet analysis. Pure Appl. Geophys. 159, 517–541.

51. Munck, F., Walgenwitz, F., Maget, P., Sauer, K., Tietze, R., 1979. Synthe` se ge´othermique du Fosse´ rhe´nan Supe´ rieur, Commission of the European Communities. BRGM Service Ge´ologique Re´gional d'Alsace, Geologisches Landesamt Baden-Wu¨ rttemberg.

52. Nami, P., Schellschmidt, R., Schindler, M., Tischner, T., 2008. Chemical Stimulation operations for reservoir development

of the deep crystalline HDR/EGS system at Soultz-sous-Fore^ts (France). In: 32nd Workshop on Geothermal Reservoir Engineering. Stanford University, California, USA 11 p.

53. Pauwels, H., Fouillac, C., Fouillac, A.-M., 1993. Chemistry and isotopes of deep geothermal saline fluids in the Upper Rhine Graben: Origin of compounds and water-rock interactions. Geochim. Cosmochim. Acta 57, 2737–2749.

54. Phillips, W.S., 2000. Precise microearthquake locations and fluid flow in the geothermal reservoir at Soultz-sous-Fore^ts, France. Bull. Seismol. Soc. Am. 90, 212–228.

55. Place, J., Naville, C., Ge´ rard A., 2007. Test of VSP efficiency for the recognition of highly dipping faults in a deep granitic basement, unpublished GEIE report, 27 p.

56. Portier, S., Vuataz, F.-D., Sanjuan, B., Nami, P., Ge´ rard, A., 2009. Chemical stimulation techniques for geothermal wells: experiments on the 3- well EGS system at Soultz-sous-Fore^ts. Geothermics doi:10.1016/ j.geothermics.2009.07.001.

57. Pribnow, D., Schellschmidt, R., 2000. Thermal tracking of Upper crustal fluid flow in the Rhine Graben. Geophys. Res. Lett. 27, 1957–1960.

58. Rummel, F., Hettkamp T., Weber, U., 1997. Density measurements on fluid samples obtained during the HDR circulation test 97July12 (Part II), Yellow Report no. 18, Ruhr-University Bochum.

59. Sanjuan, B., Millot, R., Dezayes, C., Brach, M., 2010. Main characteristics of the deep geothermal brine (5 km) at Soultz-sous-Fore^ts (France) determined using geochemical and tracer test data. C. R. Geoscience 342, 546–559.

60. Sanjuan, B., Pinault, J.-L., Rose, P., Ge´ rard, A., Brach, M., Braibant, G., Crouzet, C., Foucher, J.-C., Gautier A., Touzelet, S., 2006a. Geochemical fluid characteristics and main achievements about tracer tests at Soultz-sous-Fore^ts (France), BRGM/RP-54776-FR final report, 64.

61. Sanjuan, B., Pinault, J.-L., Rose, P., Ge´ rard, A., Brach, M., Braibant, G., Crouzet, C., Foucher, J.-C., Gautier, A., Touzelet,

S., 2006b. Tracer testing of the geothermal heat exchanger at Soultz-sous-Fore^ts (France) between 2000 and 2005. Geothermics 35, 622–653.

62. Sausse, J., Dezayes, C., Dorbath, L., Genter, A., Place, J., 2010. 3D fracture zone network at Soultz based on geological data, Image logs, microseismic events and VSP results, C. R. Geoscience 342, 531– 545.
63. Schellschmidt, R., Schulz, R., 1991. Hydrogeothermic studies in the Hot Dry Rock project at Soultz-sous-Fore^ts. Geotherm. Sci. Tech. 3, 217–238.
64. Schindler, M., Nami, P., Schellschmidt, R., Teza, D., Tischner, T., 2008. Correlation of hydraulic and seismic observations during stimulation experiments in the 5 km deep crystalline reservoir at Soultz. In: EHDRA Scientific Conference, Soultz-sous-Fore^ts. France, September 8 p.
65. Schnaebele, R., Haas, J.-O., Hoffmann, C.R., 1948. Monographie ge´ologique du champ pe´trolife` re de Pe´ chelbronn. Mem. Serv. Carte Geol. Alsace 7, 254.
66. Smith, M., 1983. A history of Hot Dry Rock geothermal systems. J. Volcanol. Geotherm. Res. 15, 1–20.
67. Stussi, J.-M., Cheilletz, J.M., Royer, J.-J., Che`vremont, P., Fe´ raud, G., 2002. The hidden monzogranite of Soultz-sous-Fore^ts (Rhine Graben, France), Mineralogy, petrology and genesis. Geol. France 1, 45–64.
68. Tester, J.W. et al., 2006. The future of geothermal energy, Impact of Enhanced Geothermal Systems on the United States in the 21st century, MIT report.
69. Valley, B., 2007. The relation between natural fracturing and stress heterogeneities in deep-seated crystalline rocks at Soultz-sous-Fore^ts (France), PhD thesis, ETH-Zu¨ rich, Switzerland, 260 p.
70. Valley, B., Evans, K.F., 2007. Stress state at Soultz-sous-Fore^ts to 5 km depth from wellbore failure and hydraulic observations. In: 32nd Workshop on Geothermal Reservoir Engineering. Stanford University, California, USA 10 p.

71. Vuataz, F.-D., Brach, M., Criaud, A., Fouillac, C., 1990. Geochemical monitoring of drilling fluids: a powerful tool to forecast and detect formation waters, SPE. Format. Eval. 177–184.

72. Ziegler, P.A., 1992. European Cenozoic rift system. Tectonophysics 208, 91–111.

Chapter 6

Thermal–hydrologic–mechanical–chemical Processes in the Evolution of Engineered Geothermal Reservoirs

Joshua Taron and Derek Elsworth

Department of Energy and Mineral Engineering and Center for Geomechanics, Geofluids, and Geohazards (G3), Pennsylvania State University, University Park, PA, USA

ABSTRACT

In a companion paper [Taron J, Elsworth D, Min K-B. Numerical simulation of thermal–hydrologic–mechanical–chemical processes in deformable, fractured porous media. Int J Rock Mech Min Sci 2009; doi:10.1016/j.ijrmms.2009.01.008] we introduced a new methodology and numerical simulator for the modeling of thermal–hydrologic–mechanical–chemical processes in dual-porosity media. In this paper we utilize the model to examine some of the dominant behaviors and permeability-altering mechanisms that may operate in naturally fractured media. Permeability and porosity are modified as fracture apertures dilate or contract under the influence of pressure solution, thermo–hydro-mechanical compaction/dilation, and mineral precipitation/dissolution. We examine a prototypical enhanced geothermal system (EGS) for the relative, temporal arrival of hydro-mechanical vs. thermo-mechanical vs. chemical changes in fluid transmission as cold (70 C) water is injected at geochemical disequilibrium within a heated reservoir (275 C). For an injection-withdrawal doublet separated by ~670 m, the results demonstrate the strong influence of mechanical effects in the short-term (several days), the influence of thermal effects in the intermediate term (<1 month at injection), and the prolonged and long-term (>1 year) influence of chemical effects, especially close to injection. In most of the reservoir, cooling enhances permeability and increases fluid circulation under pressure-drive. We observe thermo-mechanical driven permeability enhancement in front of the advancing thermal sweep, counteracted by the re-precipitation of minerals previously dissolved into the cool injection water. Near injection, calcite dissolution is capable of increasing permeability by nearly an order of magnitude, while precipitation of amorphous silica onsets more slowly and can completely counteract this increase over the very long-term (>10 years). For the reinjection of highly-silica-saturated water, amorphous silica is capable of drastic reduction in permeability close to the injection well. With combined action from all mechanisms, permeability change varies by two orders of magnitude between injection and withdrawal.

INTRODUCTION

In enhanced geothermal systems (EGS), as in reservoirs for the sequestration of CO_2, radioactive waste repositories, petroleum reservoirs, and other subsurface engineered facilities, fluid circulation is influenced in both the short- and long-term by chemical reaction and thermal-hydro-mechanical deformation. These forces operate upon temporally dynamic and spatially variable fluid transport properties, such as permeability, porosity, and fracture interconnectivity.

There are many considerations in the design of EGS, and most relate to the behavior of fractures. These include the potential for short-circuiting, the evolution of heat transfer surface area and of fluid residence time, working-fluid losses, and larger environmental concerns such as induced seismicity. Many processes are active in this regard. Some *inhibit* fluid transmission in dominant fractures, such as thin-film pressure solution [2], [3], [4], [5] and [6], mineral precipitation [7], [8] and [9], and thermo-mechanical aperture closure or creep. Others *enhance* transmission, such as shear dilation [10] and [11], mineral dissolution, force of crystallization [12], [13] and [14], and strain energy driven free-face dissolution [5] and [6], while shear and dilation on existing and growing fractures is considered the primary contributor to induced seismicity [15].

Simulating these behaviors currently requires a staged approach, with some of the behaviors encompassed in thermal–hydraulic (TH) or thermal–hydraulic–mechanical (THM) simulations, and others by thermal–hydraulic-chemical (THC) simulations. This may be adequate for many needs, but leaves out the rational and desirable linkage between chemical and mechanical behaviors. Pressure solution is by definition a chemical–mechanical process, as is free-face dissolution.

Other behaviors may, at first glance, seem adequately addressed through THM modeling, such as shear dilation and thermo-mechanical aperture closure. It is unclear, however, how such processes will be influenced by the action of chemical or chemo-mechanical mechanisms. For instance, aperture changes due to pressure solution result in the growth of contact area between fracture planes, and this in turn affects fracture shear strength and dilation angle. Alternatively, mineral precipitation/dissolution can alter the apertures of fractures, causing shifts in fluid and thermal flow paths that upset the stress condition and modify mechanical aperture. To examine these and other linkages, we introduce a modeling structure [1]that couples chemical and mechanical behaviors in a manner that reflects their interdependence on thermal and hydraulic effects (THMC). We apply this simulator to address some of the important questions in a prototypical EGS and, indeed, to examine the importance of these couplings and the strength of their interaction.

We first provide a brief overview of the approach and a description of the geological domain. A description of the chemical system and a simplified examination of injection water chemistry then follow to set the stage for further analysis. This is followed by an assessment of the relative, temporal arrival of hydro-mechanical vs. thermo-mechanical vs. chemical changes in fluid transmission and a recast of these in terms of characteristic times. Next, and perhaps most importantly, we evaluate the necessity for the explicit coupling of reactive transport to geomechanical processes and quantify the strength of coupling between independent processes. Finally, we follow the evolution of deviatoric stress in the reservoir and indicate locations most likely to undergo shearing failure.

SIMULATION MECHANISM

The simulations presented in the following utilize a newly developed THMC simulator [1] that couples the multiphase, multi-component, non-isothermal thermodynamics, reactive transport, and chemical precipitation/dissolution capabilities of TOUGHREACT [7] with the stress/deformation analyses FLAC^{3D}[16]. The coupled model incorporates features unique to fractured reservoirs (particularly those under elevated temperature and chemical potential), involving the undrained pressure response in a dual-porosity medium and with chemo-mechanical effects on deformation and on transport.

This "modular" approach, first proposed by Settari [17] to couple geomechanics with reservoir flow simulation, has some advantages over the development of a single coupled program. Modular approaches take advantage of the high state of development, sophistication, and validation available in purpose-build codes. The appropriate linking of codes which represent different behaviors—in this case those of solid mechanics and fluid transport—enables complex coupled processes to be represented and their interaction explored. With the appropriate choice of linking parameters and with the use of flexibly defined mesh overlays, the interaction of complex coupled processes may be explored with reasonable confidence and with rapid development.

Additionally, the modular construction [17] and [18] allows for easier implementation of future advances in constitutive or modeling technology (rather than modifying an entire coding structure), although typically the codes are not optimized to perform interactively, and may execute slowly; in our experience lowering computational efficiency due to the time required for data transfer between modules. As suggested by Settari and Mourits [18] and Minkoff et al. [19], however, this may not always be the case, because in systems where geomechanics may be loosely coupled (not changing at a rapid pace) the geomechanics simulation may not need to be conducted very often, thus improving computational efficiency over fully coupled codes where mechanics are equilibrated at every fluid flow time step.

FLAC^{3D} is exercised purely in mechanical mode, where undrained fluid pressures may be evaluated (externally) from local total stresses. This undrained methodology allows calculation of the short-time build-up in fluid pressures that results from an instantaneous change in stress, provided we have knowledge of the compressibility of the pore fluids and the solid matrix. In this way, the complex thermodynamics of phase equilibria of multiphase water mixtures, and even multi-component mixtures (such as CO_2 and water), can be tracked in the pre-existing framework of TOUGHREACT. For water mixtures, we utilize the 1997 International Association for the Properties of Water and Steam (IAPWS) steam table equations [20]to calculate fluid compressibility. If a system is unsaturated (such as in HDR) fluid compressibility is large, and the undrained poroelastic equations approach their drained counterparts. Therefore, while our construct is tailored to saturated systems, drained ones are automatically accommodated.

Sequential execution of the two programs is linked by a separate code, referred to as the "interpolation module", capable of parsing data outputs from each primary simulator as input to the companion. This module is a Fortran 90 executable, and maintains access to data outputs from TOUGHREACT and FLAC^{3D}. In addition to data interpolation, the module executes constitutive relationships including permeability evolution, dual-porosity poroelastic response to stress, and thermodynamically controlled fluid compressibility.

THMC PERMEABILITY CHANGES

The modeling structure calculates permeability change from the combined action of pressure solution, thermo-mechanical dilation/contraction, hydro-mechanical dilation/compaction, and bulk volume precipitation/dissolution of mineral species, with each depending intrinsically on temperature, effective stress, and chemical potential. Several methodologies could be used to model these behaviors; in the following we utilize laboratory results of fracture behavior under hydrothermal conditions to constrain in a single constitutive relationship the first three mechanisms. Compaction of fracture asperities is governed by [21]

$$b_{mc} = b_m^r + \{b_m^r - b_c^r + b_m^{\max} \exp(-\omega\sigma') \cdot \exp(-\sigma'(\beta - \chi/T)), \tag{1}$$

where b_{mc} is the fracture aperture at a given temperature, T, and effective stress, σ'. The residual (minimum possible) aperture at extreme mechanical stress is given by b_m^r, and at extreme chemical stress by b_c^r. The maximum possible aperture change is b_m^{max} (or the difference between initial and residual aperture), and the empirical coefficients that must be fit to laboratory data are β, χ, and ω, and represent the chemical, thermal, and mechanical dependence of the aperture, respectively (with ωrepresenting non-linear fracture stiffness). This relationship has been previously compared to laboratory data [1] and [21].

Unloading (gaping) of the fracture is controlled by [1] and [21]

$$b_{m(u)} = b_{mc}(\sigma'_{\max}) + R_m b_m^{\max}\{\exp(-\omega\sigma') - \exp(-\omega\sigma'_{\max})\}, \quad (2)$$

where $b_{mc}(\sigma'_{max})$ is the aperture at maximum loading (Eq. (1) evaluated at the maximum stress prior to unloading), and R_m is the recovery ratio (fraction of *mechanical* (not chemical) closure that is recoverable through mechanical–thermal unloading). This relationship implies that gaping of the fracture can only occur through mechanical–thermal elastic relaxation, and not through the chemical processes that influence closure in Eq. (1). The values of these parameters as utilized in our simulations appear inTable 1.

Table 1: Solid medium properties as used in simulations

Parameter	**Value**	**Parameter**	**Value**
Bulk modulus of intact rock, $K^{(1)}$ (GPa)	17.0	Volume fraction of fractures (of total reservoir), ϑ	0.01
Composite bulk modulus, *K* (GPa)	8.00	Porosity within fractures, φ	0.3
Poisson's ratio, υ	0.22	Residual mechanical aperture, b^r_m (μm)	22.5
Bulk modulus of solid grains, $K^{(1)}_s$ (GPa)	54.5	Residual chemical aperture, b^r_c (μm)	20.0
Coefficient thermal expansion, α_T (1/°C)	1.2×10^{-5}	Aperture change constant, β	1.11
Saturated thermal conductivity, λ (W/m K)	2.9	Aperture change constant, χ	345

Heat capacity, C_p (J/kg K)	918	Aperture stiffness constant, ω (1/MPa)	0.20
Porosity of porous media, φ	0.02	Mechanical recovery ratio, R_m	0.60

Chemical precipitation/dissolution is calculated within TOUGHREACT. The governing equations are described in Taron et al. [1], as presented in the works of Steefel and Lasaga [22] and Xu et al. [23]. Reactions between aqueous species (homogeneous reactions) are assumed to be at local equilibrium, and therefore governed by the relationship between the concentrations of basis (primary) species and their activities, partitioned by the stoichiometric coefficients. This relationship is termed the law of mass action[cf. 22]. The assumption of local equilibrium greatly reduces the number of chemical unknowns and ODE's (between primary and secondary species), and is accurate to the extent that the true reaction rates outpace the rate of fluid transport in a given system. This is a correct assumption for most aqueous species [22] (and flow systems), but less so for slower redox reactions [22] and [23]. In TOUGHREACT, species activities are obtained from an extended *Debye–Hückel* equation with parameters from Helgeson et al.[24].

MODEL RESULTS

In the engineered stimulation of geothermal reservoirs hydraulic, chemical [8], [9], [25] and [26], and thermal [27], [28], [29] and [30] means may be utilized to enhance permeability in a purposeful manner. These stimulations may be used to increase circulation in low permeability systems, and also to counteract future changes that occur as cold water is injected into the reservoir at geochemical disequilibrium. Ideally, injection and withdrawal wells will be connected by a fracture network that is sufficiently permeable to allow rapid fluid interchange, but impermeable enough to allow for sufficiently long thermal residence times in a medium with a large heat transfer area. In addition to achieving an appropriate balance between circulation and residence time, a natural concern arises as to whether unintended stimulations (resulting from critical thermo–hydro-chemical forces of injection) will generate short-circuiting paths between the injection and withdrawal wells via the modification of flow pathways through existing fractures.

Mineral scaling at the wellbore is a known problem, and recent work [25] and [31] indicates that mineral precipitation may also play a large role in reservoir evolution. Moderate success (primarily for calcite) has been achieved in counteracting this through pH modification and the addition of chelatants [8] and [9]. Chemical changes and the effectiveness of chemical treatments, however, are coupled to hydro-mechanical fracture flow properties. At the Soulz-sous-Forêts geothermal site, for instance, productivity enhancement was observed from chemical treatment at two production wells, but had almost null impact at the injection well, presumably due to the presence of high conductivity fractures that transmitted the treatments quickly away from the problem area [9] and ameliorated their impact. The purpose of this work is to explore the chemical–mechanical couplings responsible for such behaviors and to quantify how strongly they are interlinked.

Characteristics of Simulated EGS

We examine behavior in a prototypical EGS represented by a pseudo 3D doublet geometry of 1500 m×2900 m, with cold (70 °C) fluid injected through a single well and extracted (670 m away) through a single withdrawal well (Fig. 1). Initially, the reservoir is at 275 °C. Horizontal stresses of 38.9 MPa and a vertical stress of 61.7 MPa are applied at time t=0^{-} and allowed to equilibrate with an initial pore pressure of 24 MPa. These conditions are broadly representative of the east flank of Coso geothermal field [32] and [33]. The lateral boundary that crosses both the injection and withdrawal wells is a plane of symmetry, and as such is restricted to zero displacement in its orthogonal direction (roller boundary). Constant normal stress is held at all other boundaries, and every boundary is thermally insulated and impermeable. The fracture domain is given an initial (unstressed) permeability of 3.0×10^{-11} m^2, and the application of effective stress and temperature at $t=0^-$ are allowed to decrease this permeability to an *in-situ*, equilibrium value via mechanical closure and pressure solution (assumed to have occurred slowly over geologic time prior to reservoir stimulation). This *in-situ* value is 6.8×10^{-15} m^2 at the given conditions. The granodiorite matrix is given a permeability of 2.0×10^{-18} m^2. Additional properties of the solid medium are shown in Table 1. The injection well operates as a constant source of pressure and temperature equivalent to a well diameter of approximately 12 in, while the withdrawal source serves as a single withdrawal well maintaining a constant pressure at the same diameter. The pressure differential is 3 MPa above and below the *in-situ* value, respectively, and both enter the geometry at the boundary of symmetry.

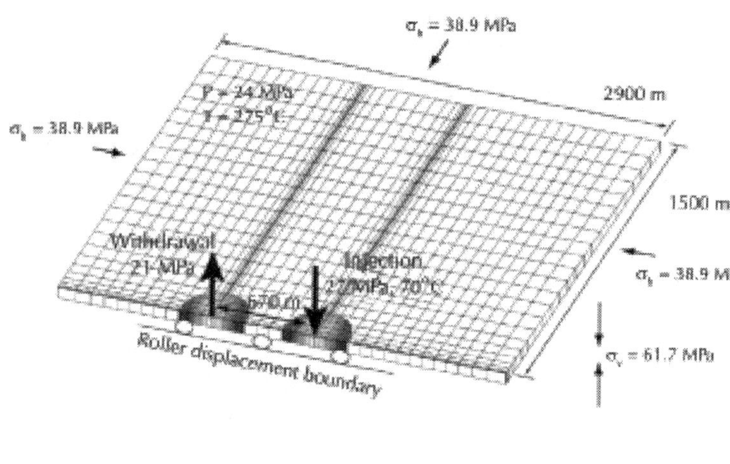

Figure 1: Initial conditions and geometric layout of EGS reservoir as used in simulations.

Reactive composition of the host reservoir rock is presented in Table 2. Calcite and amorphous silica are expected to be the minerals primarily responsible for permeability change due to precipitation and dissolution [25] and [33]. Other likely minerals, such as potassium feldspar and quartz, are also followed, as listed in Table 2. Rate constants for precipitation/dissolution and mineral reactive surface areas of these common minerals are available in the literature, and were utilized as in Xu and Pruess [25]. Calcite and anhydrite are assumed to react at equilibrium (benefiting simulation time by lowering the *Damköhler*number). This assumption is warranted by their very fast relative rate of reaction.

Table 2: Initial volume fraction of reactive minerals in host reservoir

Mineral	**Volume fraction of solid rock[a]**	
	Granodiorite	Fractured vein
Anhydrite	–	–
Anorthite	0.33	–
Calcite	0.02	0.31
Chlorite	–	0.23
K-feldspar	0.17	–
Quartz	0.34	0.17
Amorphous silica	–	–

Remaining volume fraction (up to 1) is non-reactive.
[a]From [33].

Mineral Precipitation/Dissolution

Fluids that are injected into a geothermal reservoir are typically far from geochemical and thermal equilibrium with minerals in the host rock. As minerals precipitate from the injected fluid and dissolve from the rock, changes are incurred in the porosity and permeability of the fracture system. In currently operating geothermal systems, calcite and amorphous silica precipitation have posed problems at recovery and injection wells, respectively. Acidic injection has been shown to successfully inhibit calcite precipitation [8] and [15]. Aggressive acid–base reactions, however, tend to rapidly dissolve first-contacted minerals and lose effectiveness in the remainder of the wellbore [8]. Other problems with acidification include the potential for corrosion of steel casings, although corrosion inhibitors can often be used. Chelating agents, which reduce the activity of metal ions (through binding), are another alternative for calcite dissolution [9] and [34] that are suggested to have a less aggressive, more uniform effect [8] and [34].

Inhibiting the precipitation of amorphous silica has proven more difficult [15]. HF treatment is currently the preferred method, but some have suggested that amorphous silica can be dissolved at high pH with the use of chelating agents to prevent calcite deposition (which would be favored at high pH) [8] and [34]. This research is, however, relatively recent, and uncertainty remains as to the effects of these treatments on the host rock. Geothermal environments are chemically complex, and field scale results are limited and mixed [9].

In the simulations that follow we present three scenarios for injection fluid composition. Injection water #1 is obtained by extracting the equilibrium, *in-situ* reservoir fluid at 275 °C and allowing it to cool (not in the presence of reactive minerals, as if utilized at the surface with no treatment other than settling) to 70 °C while the aqueous components equilibrate and minerals precipitate. For this injection water, our results show that the injection fluid is not sufficiently saturated with aqueous silica to incur significant reservoir precipitation of amorphous silica, although other minerals are very active. This is a result of the chosen chemistry. As a comparison, field measurements have shown that the precipitation of amorphous silica occurs in fractures at some injection locations in the Coso geothermal field (well 68-20RD), but not at others (wells 68-20, 68A-20, 68A-20RD) [35], depending on the precise injection and *in-situ* chemistry at each well. Two alternative compositions are introduced to examine the behavior of amorphous silica precipitation. Injection water #2 assumes some inefficiency in the surface equilibration process (such as insufficient equilibration time or enhanced dissolution from turbulence) such that only 80% of potential silica precipitation occurs. Simulation #3 is further oversaturated in silica; it represents the potential for rapid chemical change in the reservoir and so assumes that mineral precipitation has been completely restricted during the cool down from 275 to 70 °C (only precipitation is restricted, as primary species are allowed to form secondary species in solution). Injection compositions for each of these cases and for the*in-situ* reservoir are listed in Table 3.

Table 3: Fluid chemical compositions (mol/l) used in simulations

Chemical component	**Reservoir fluid**	**Injection fluid**		
		Water #1: initial fracture vein water cooled and equilibrated at 70 °C	Water #2: allowing only 80% completion of SiO_2 precipitation during equilibration to 70 °C	Water #3: restricting all mineral precipitation during equilibration to 70 °C
Al	8.520×10^{-3}	1.894×10^{-8}	1.894×10^{-8}	3.866×10^{-3}
Ca	1.020×10^{0}	1.976×10^{0}	1.976×10^{0}	1.976×10^{0}
Cl	7.420×10^{1}	1.074×10^{2}	1.074×10^{2}	1.074×10^{2}
Fe	4.531×10^{-9}	8.167×10^{-8}	8.167×10^{-8}	8.167×10^{-8}
HCO_3	2.972×10^{-2}	2.010×10^{-1}	2.009×10^{-1}	2.010×10^{-1}
K	6.136×10^{0}	8.413×10^{0}	8.413×10^{0}	8.435×10^{0}
Mg	1.213×10^{-3}	3.938×10^{-3}	3.938×10^{-3}	3.938×10^{-3}
Na	6.359×10^{1}	9.255×10^{1}	9.255×10^{1}	9.255×10^{1}
$SiO_{2(aq)}$	6.210×10^{0}	1.549×10^{0}	4.881×10^{0}	7.856×10^{0}
SO_4	2.766×10^{-2}	7.193×10^{-2}	7.193×10^{-2}	7.193×10^{-2}
pH	6.71	5.82	5.82	5.82

Reservoir composition is at time, *t*=0. Injection fluid composition remains constant throughout simulation.

The mineral precipitation/dissolution results of our simulations utilizing these three injection waters are shown in Fig. 2. Fig. 2A compares each injection water for a simulation time of 5 and 20 years for the combined action of all minerals, while Fig. 2B illustrates results for injection water #2 and for each mineral individually. Units are cumulative change in mineral abundance (volume fraction) from time, *t*=0, such that negative values represent dissolution. Values in Fig. 2B are normalized, such that multiplication of each curve by its normalization magnitude returns the units to cumulative change in mineral abundance. Importantly, water #1 exhibits net mineral dissolution near the injection well (where the cold injection water makes calcite dissolution highly favorable), and net precipitation outside of the thermal reach. This is desirable in the sense that the immediate injection area experiences an increase in permeability, but undesirable in that these dissolved minerals are then re-precipitated as they are heated again between the injection and withdrawal wells. Water #2 changes this behavior in a crucial manner; at 5 years the injection area still experiences net dissolution, but this magnitude is decreased by the precipitation of amorphous silica, and at 20 years the action of amorphous silica has completely erased the positive influence of calcite dissolution. Water #3 shows a more dramatic result, where silica precipitation completely dominates the injection area. This behavior is so pronounced that the simulation was ceased after 7 years due to a complete shutdown of injection permeability ($<10^{-18}$ m^2), and so only the 5 year value is plotted in the figure. One final point regards the magnitude of mineral precipitation outside of the injection area. The higher hydraulic gradient between injection and withdrawal partially outpaces mineral precipitation, such that there is a higher buildup of deposited minerals on the opposite side of injection due to a lower fluid velocity (higher *Damköhler* number).

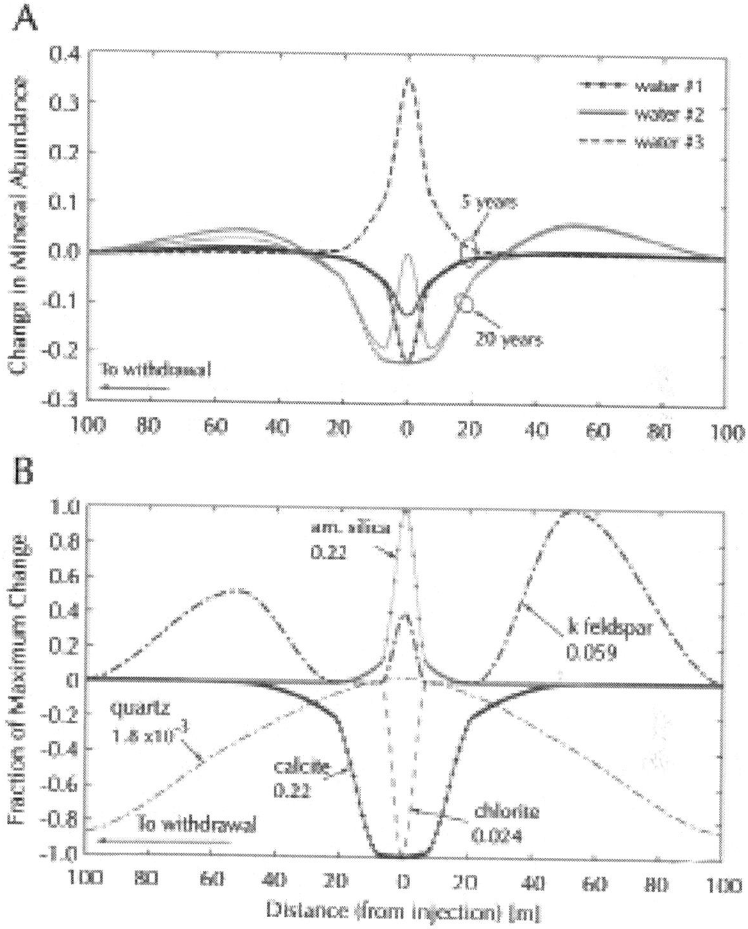

Figure 2: (A) Change in mineral abundance (all minerals combined) at 5 and 20 years for each of the three injection compositions. Negative values indicate dissolution (decrease in the solid volume fraction) and (B) behavior of each mineral utilizing water #2. Each curve normalized to its own maximum value: Number beneath each mineral indicates the normalizing value (multiply by this number to obtain the true value of each curve).

Further parametric analyses of injection chemistry are not addressed here [7], [8], [33] and [35]. The primary interest in what follows is to present an analysis *not* of specific chemical changes, but of how these changes impact the mechanical system, and how feedback from other permeability altering mechanisms may shift chemical behaviors. We begin by examining the competing magnitudes of chemical vs. hydraulic vs. thermal effects.

Temporal Permeability Change

Our model so far attempts to interlink the most important mechanisms in the alteration of reservoir fluid transport properties. This section reverses direction and attempts to extract from the coupled analysis the magnitude of influence for each process. It is of interest, both for overall reservoir evolution and for evaluation of potential hydraulic/chemical stimulations, to observe the temporal onset of the various mechanistic causes: In what stages of the lifespan of a geothermal project is thermal (or chemical or hydraulic) forcing the dominant mechanism for change? What measures could then be taken to counteract this, and when should they be applied?

A conceptual model for the possible temporal relationships between THMC processes is shown in Fig. 3A. In a geothermal reservoir subject to cold water injection, hydraulic stimulation should have an immediate effect, tapering off with time to a steady long-term influence. Thermal effects would exhibit a similar trend, but the thermal transfer rate would greatly slow the onset of changes, and there would perhaps be slight reversibility following passage of the thermal front (as adjacent grid blocks contract). Chemical effects should act even more slowly, and perhaps have a dominant influence in the extreme long-term of reservoir life. Mechanical effects are directly active in the first two of these processes—through effective stress and thermal stress.

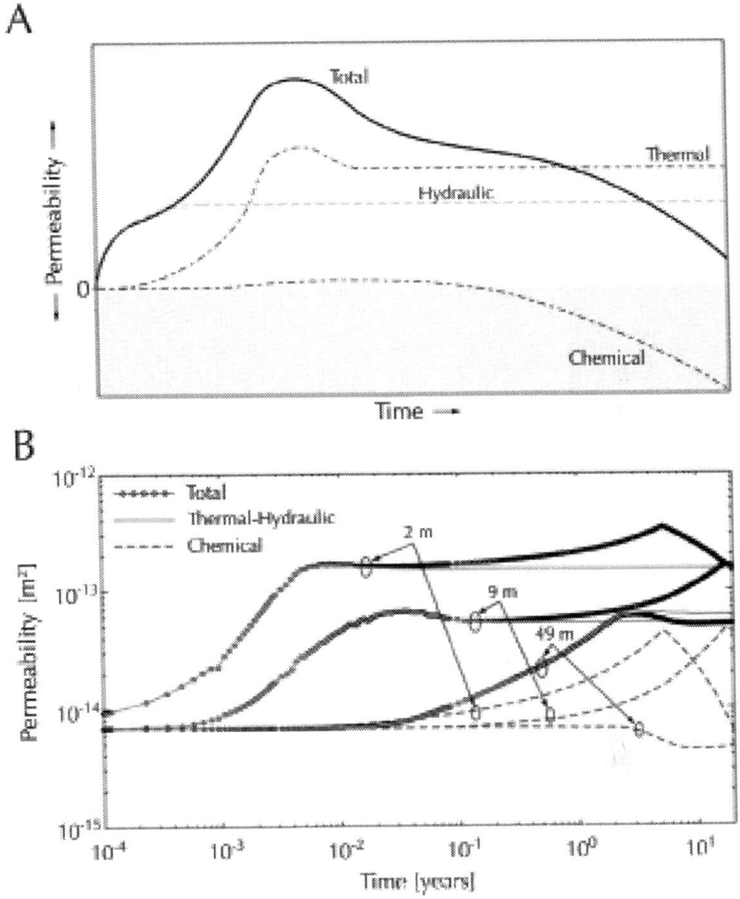

Figure 3: (A) Conceptual model of permeability altering processes after [26] and (B) contribution of each mechanism to total permeability change in our simulations. Analogous to (A). Changes monitored at 2, 9, and 49 m from injection. Uses water #2.

Simulation results are presented in Fig. 3B in a manner analogous to the conceptual model of Fig. 3A. Aperture change is monitored at three locations in the reservoir (2, 9, and 49 m outward from injection), and each curve at each location represents the contribution of a different process. Total aperture change refers to the contribution from all mechanisms and so reflects the observable change in permeability. Also shown is the contribution to total permeability change due to mineral precipitation/dissolution and TH processes (asperity dissolution via pressure solution and thermal gaping due to cooling). For asperity dissolution/thermal gaping, behavior is conditioned by the combined influence of effective (not total) and thermal stress, and so in this analysis hydraulic and thermal effects cannot be separated (although we present some separation in terms of characteristic times in the following section).

Some clarification is helpful in regard to the two processes implied by TH. In our pressure solution model (see Section 3 and Taron et al. [1]), fracture gaping is not allowed via chemical means ("force of crystallization" is not considered), but only occurs through thermo-mechanical-hydraulic unloading. This behavior, however, is not simply calculated via thermal shrinkage of the matrix blocks, but is handled constitutively as a function of stress reduction and temperature change at the fracture face, and is therefore calculated from the empirical closure relationship (Eqs. (1) and (2)). When the reservoir is loaded, pressure solution is active and instantly (numerically) reduces the reservoir permeability to its "initial" state at loaded conditions (assumed to have occurred slowly over geologic time before stimulation of the reservoir). As the reservoir cools, the matrix contracts and most areas of the reservoir experience thermo-mechanical gaping, while some areas experience an increase in stress and a continuation of pressure solution creep. Both of these actions are controlled by the constitutive relationship, and because thermal gaping is the dominant mechanism during cooling, we refer to both with the TH curves in Fig. 3B.

Mineral precipitation/dissolution is represented by the dashed lines in Fig. 3B. At points near injection, permeability is increased significantly in early times as calcite dissolves into the cool injection water; this behavior is not visible at 49 m from injection. During this time, amorphous silica precipitation counters the dissolution of calcite and decreases permeability, but this behavior is not visible in the plot due to strong dominance from calcite dissolution. As time proceeds, silica is more successful at decreasing permeability and after 5.2 years (2 m from injection) and 17.5 years (9 m from injection) nearly all of the existing calcite within fractures has been dissolved and amorphous silica overrides as a net decreasing trend in permeability. Note that in this simulation calcite dissolves at equilibrium, and that the introduction of kinetic constraints could partially shift the behavior of calcite. At 49 m from injection, only precipitation is significant during the first 20 years of simulation. Note also that neither anhydrite or anorthite is shown in the figure; their contribution is smaller by several orders of magnitude and is focused on the edge of the thermal front 100 m from injection.

Fig. 4 shows contours of permeability and temperature for a reservoir plan-view after 20 years of simulation. Note that the highest permeability is at the injection well, and this tapers off to either side and is surrounded radially by a pocket of lower permeability, particularly to the opposite side of withdrawal. Fig. 5shows permeability change at a cross section between wells. Note the nearly two order of magnitude variability, and the development of a radial barrier of lower permeability approximately 50 m outward from injection. Fig. 5B illustrates more clearly how this barrier corresponds to the re-precipitation of minerals that are dissolved at injection.

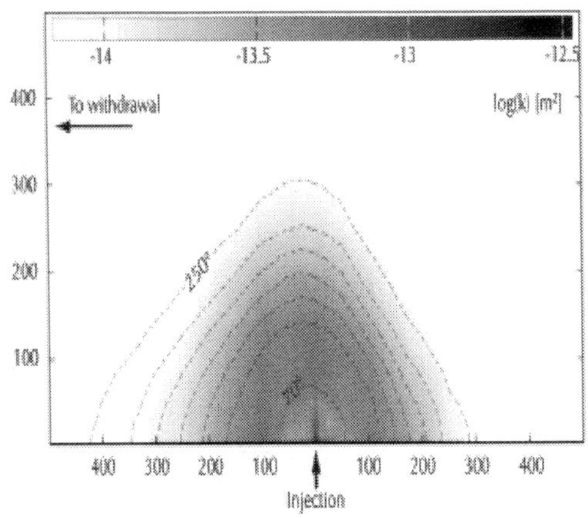

Figure 4: Plan-view distribution of logarithm of permeability (filled contours) and isotherms (empty dashed contours) at 20 years using water #2. Isotherms increment by 30 °C from 70 to 250 °C.

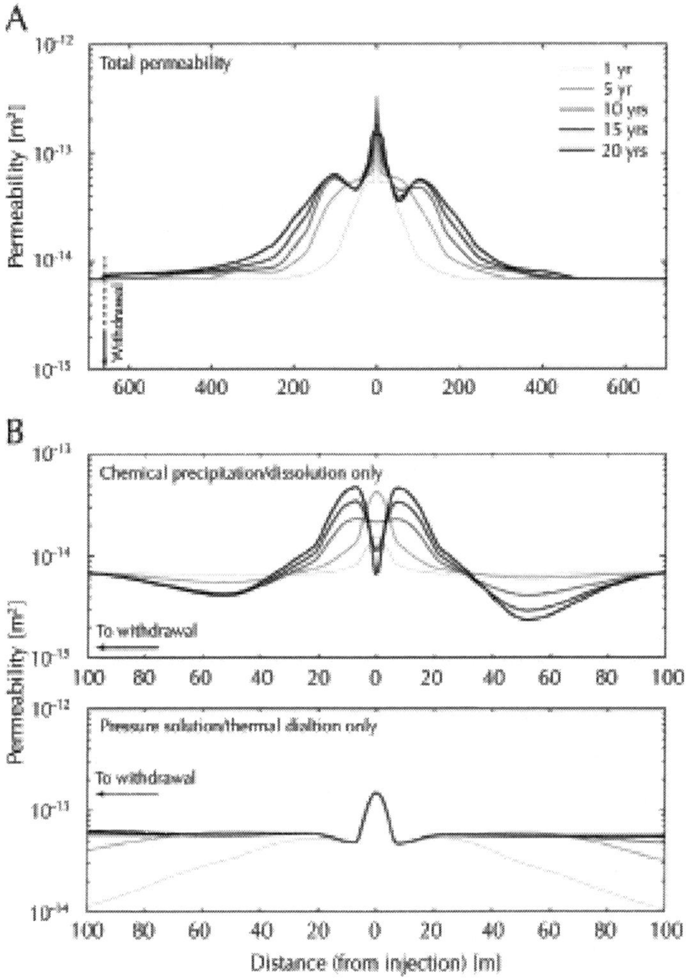

Figure 5: Permeability changes using water #2: (A) total permeability change with distance from injection and (B) mineral precipitation/dissolution portion and thermal–hydraulic portion of total permeability change in (A). Note the change of scale on thex-axis of (B).

Characteristic Times

We can further elucidate the time dependency of competing mechanisms by plotting them vs. their respective characteristic times.

We thus introduce the characteristic time, $t_D=ct/l^2$, at a time, t, over the characteristic length scale, l. The diffusivity, c, refers to the ratio of conductivity to capacity/storage, and for thermal conductance is represented by thermal conductivity of saturated rock, , density, , and heat capacity, C_p, as $c_T= / C_p$. When examining the characteristic thermo-elastic response of the reservoir, we note that the timescale of thermal expansion due to a change in temperature will be nearly instantaneous in comparison to the rate of thermal/hydraulic transfer that allows such a temperature change. Hydraulic transport will also influence the thermal characteristic time, such that cooler fluid will be transported via the fracture network to the characteristic location before being thermally conducted into the rock matrix. We conceptualize this process as two conductors in series; thermal characteristic time is recast as $_{tD}$=t·$_{cTH}$, where c_{TH} is thermo-hydraulic diffusivity and includes hydraulic and thermal effects as $1/C_{TH} = l_T^2/C_T + l_T^2/C_T + l_H^2/C_H$, where the thermal characteristic length scale, l_T, is fracture spacing and the hydraulic length scale, l_H, is the distance of the characteristic location from the injection well. Hydraulic diffusivity, c_H, is discussed below.

Hydraulic effects are characterized by the poroelastic fluid diffusivity, $c_H=k/\mu S$, for permeability, k, dynamic viscosity, μ, and 3D storage coefficient, S (see Wang [36] page 55 for a discussion of storage coefficients), defined as

$$S = \frac{1}{K_v} + \frac{\phi}{K_f}, \tag{3}$$

for the effective porosity of the fractured medium, , fluid bulk modulus, K_f, and for the solid modulus, $K_v=K+4/3G$, where K and G refer to the bulk and shear modulus, respectively. Again, the solid response to pressure change is much faster than the rate of fluid conductance, and so the hydraulic poroelastic response is characterized by this hydraulic diffusivity.

For chemical effects, we introduce an analogous relationship. Whereas before we were interested in the ratio of transport to storage, chemical effects depend on the relationship between the rate of chemical reaction and the storage capacity of chemical species in the fracture system. This gives the chemical characteristic time as the ratio of chemical reaction rate to volume capacity of the fractures,

$$t_D = \frac{k_R S_A}{(\theta\phi V/V_M)} \cdot t, \tag{4}$$

for the reaction rate, k_R (in mol/m/s)), total area of precipitation/dissolution (fracture surface area), S_A, volume fraction of fractures in the reservoir, ϑ, porosity of these fractures, φ, characteristic volume scale,V, and molar volume of the mineral species, V_M. We can also rearrange this relationship to produce a "chemical diffusivity" analogous to the above relationships, $c_C=k_R S_A/(\vartheta\varphi h/V_M)$, for reservoir width in thez-direction, h, so that the characteristic time retains its form as $t_D=c_C t/A$, for the characteristic reservoir area, $A = \pi l_c^2$, for the characteristic length, l_c, which is the distance of the characteristic location from the injection well. We thus obtain three characteristic times (and diffusivities) to represent the thermal, hydraulic, and chemical aperture response.

In Fig. 6, four characteristic areas are examined within the influence field of injection; one of 9 m characteristic length (encompassing the area closer than 9 m from injection), one of 24 m, one of 49 m, and one of 129 m, and we take 70 °C (corresponding to the water injection temperature) to be the dominant temperature condition for the life of the project (within these characteristic areas). Water properties correspond to this temperature, and reaction rate is adjusted to this temperature via the Arrhenius expression,

$$k_R = k_{25} \exp\left(-\frac{E_a}{R_u}\left(\frac{1}{T} - \frac{1}{298.15}\right)\right), \tag{5}$$

for the gas constant, R_u, activation energy, E_a, and the rate constant at 25 °C, k_{25}.

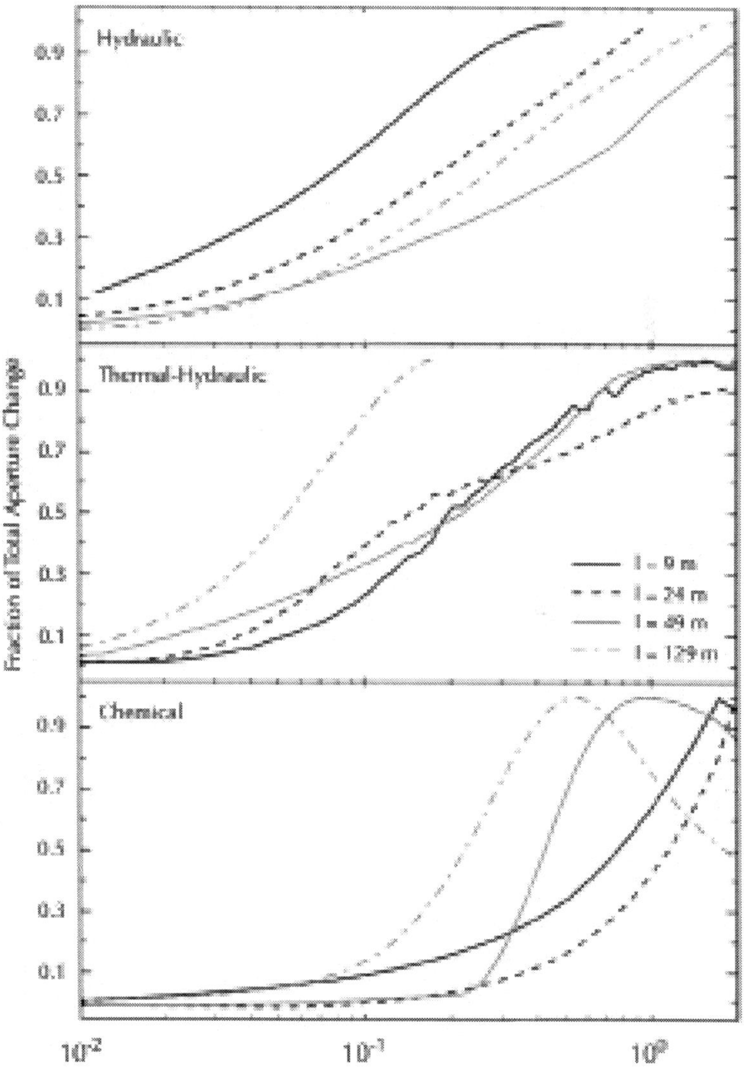

Figure 6: Characteristic times of permeability change due to hydraulic, thermal–hydraulic, and chemical mechanisms. Changes monitored at 9, 24, 49, and 129 m from injection. Uses water #2.

As mentioned above, asperity dissolution/thermal gaping behavior is conditioned by effective (not total) and thermal stress and so hydraulic and thermal effects are not directly separable. However, we can focus on hydraulic effects by extracting from the simulation data any changes that occur prior to any noticeable change in temperature; because hydraulic effects are deduced to occur well in advance of thermal arrival. This pre-thermal onset time varies from approximately 10 minutes at 9 m from injection to 4 days at 129 m from injection. Closer than 9 m from injection we are unable to distinguish between mechanisms. Fig. 6separates hydraulic and thermal effects (both contributing to the empirical aperture change relationship) via this observation, so that data points prior to this shift in behavior are assumed to relate to hydraulic influence (and become the hydraulic curve in Fig. 6). Because the thermal–mechanical relationship includes the effect of hydraulic diffusivity (see above), no separation is required here and the curve is plotted from time, t=0 (this is the true TH curve). To examine characteristic chemical behavior, we adopt for simplicity the rate constant and activation energy of amorphous silica precipitation.

From Fig. 6, we note that for the three plots (hydraulic, TH, and chemical), most of the changes fall characteristically within a timescale of t_D=1.2. Ideally, a curve at any location would overlay perfectly with any other location by characteristic scaling. Given the strong couplings expected between each of the three mechanisms, however, it is perhaps surprising to observe any tendency to follow this ideal, particularly for the TH curves.

In addition to providing a general picture of characteristic reservoir changes, this analysis would seem to have implications for the strength of coupling between each mechanism: a perfect overlay of all curves could indicate a weak coupling. For instance, that TH effects do not significantly influence chemical trends, such that reactive transport simulations at invariant stress would not suffer a penalty in accuracy for this assumption. The following section presents a more direct analysis of this coupling strength.

Strength of Coupling

The primary justification for a THMC modeling strategy is that there exists a strong coupling between chemical and mechanical effects that cannot be represented by parallel reactive transport and geomechanics simulations. In this section we examine the truth of this assertion. We first note that pressure solution can only be simulated with chemo-mechanical coupling, and this alone provides partial justification. However, in the current equilibrium based pressure solution model, coupling to a reactive transport framework is not required (such a linkage may be desirable, but its necessity is currently unclear). Further justification of coupling strength is desirable.

By examination of Fig. 3, Fig. 4 and Fig. 5 it is clear that both reactive transport and geomechanical (THMC asperity dissolution/dilation) analysis are required to reproduce the observed order of magnitude changes in permeability; with each causing changes at least as large as one order of magnitude. Furthermore, our pressure solution relationship reduces initial permeability from 3×10^{-11} to 6.8×10^{-15} m^2 at *in-situ* reservoir conditions (this initial drop is not shown in the figures). Therefore, both of these methodologies are important, and both are required to follow reservoir behavior under these prototypical EGS conditions; but this conclusion does not necessarily imply that they must be interlinked.

Fig. 7 examines this necessity. Fig 6A and B show permeability changes *due only to mineral precipitation/dissolution* (from an initial value of 6.8×10^{-15} m^2). Two different line styles represent two different simulations; the first utilizes our coupled THMC scheme, and the second operates under invariant stress (no mechanical equilibration and thus no pressure solution or thermal–mechanical dilation). Therefore, differences between the two simulations indicate shifts in mineral precipitation/dissolution behavior that are *caused by changes in the mechanical system*. In Fig. 7A, at 50 m to the left of injection, a reactive transport simulation shows that mineral behaviors cause a reduction in permeability from 6.8×10^{-15} to 5.4×10^{-15} m^2, while the coupled THMC simulation shows a drop to 2.3×10^{-15} m^2, or a 3-fold increase in the amount of permeability occlusion, while 10 m to the left of injection shows a 3-fold increase in permeability enhancement. At injection, there is further evidence of this, where amorphous silica precipitates more rapidly in the THMC coupling, resulting in a lower final permeability. Fig. 7B plots these same simulations vs. time, and again illustrates the strong dependence of mineral reactions on mechanical changes.

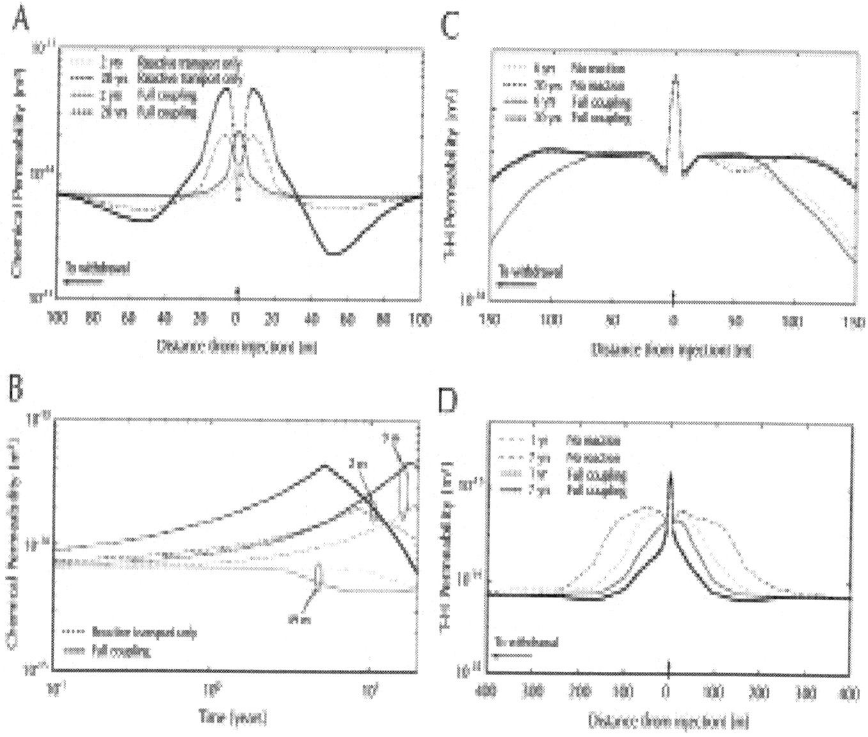

Figure 7: Permeability comparisons for degrees of coupling: (A) permeability changing due to mineral precipitation/dissolution only: compares full coupling vs. reactive transport alone (invariant stress, no stress dependent aperture change), (B) same as (A) vs. time, (C) permeability changing due to pressure solution/thermal dilation only: compares full coupling vs. exclusion of reactive transport, and (D) same as (D), using water #3 (all others use water #2).

Fig. 7C and D invert the comparison to examine how changes in the chemical system may alter the evolution of mechanics. Here we plot permeability change *resulting only from* pressure solution/thermo-mechanical gaping, with the same initial value as before. The lines again represent two simulations; one THMC coupled, and a second that does not allow chemical reaction. Therefore, differences between the two simulations indicate shifts in the thermal–mechanical system *caused by changes in the chemical system.* In Fig. 7C we again see shifts in pressure solution/thermal dilation that are caused by precipitation/dissolution, but the coupling is not as strong as for the inverse condition in Fig. 7A. To examine the potential for greater coupling, Fig. 7D conducts the same simulation but utilizes water #3 (more chemically rich). In this case, very large, order of magnitude differences in thermal–mechanical dilation occur when chemical reaction is included.

Evolution of Stress

Deviatoric stress that arises around thermal input can alter fluid circulation and generate a microseismic response. When the deviatoric stress exceeds the bulk strength of reservoir rock or existing fracture planes, permeability may be increased through the dilation and/or extension of existing fractures. This will alter fluid flow paths, and introduce the potential for microseismicity. Fig. 8 follows the advance of deviatoric stress ($\frac{1}{2}(\sigma_1 - \sigma_3)$) with shaded contours. Included on the plot are white vectors to indicate locations where shear stress currently exceeds the reservoir shear strength at that location, and they scale with the excess stress to strength ratio. Their direction indicates the likely direction of failure were it to occur (orthogonal to minimum principal stress). Shear strength is calculated based upon an internal angle of friction of 25° and zero cohesion.

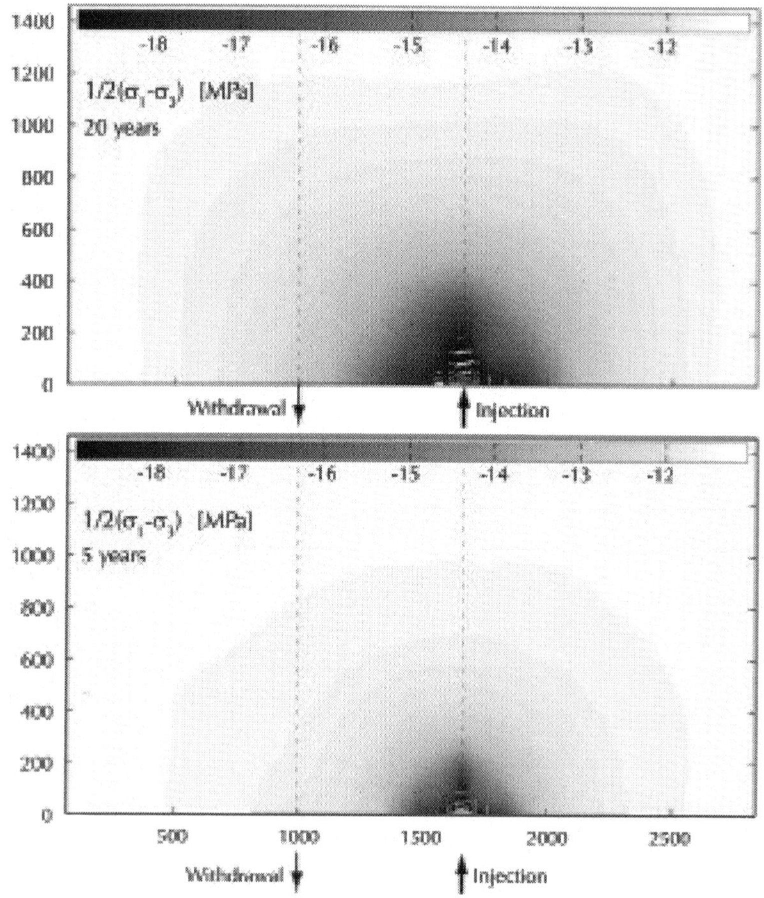

Figure 8: Plan-view of reservoir with deviatoric stress contours (in MPa) at 5 and 20 years of simulation. White lines are oriented as orthogonal to minimum principal stress direction, and are scaled to the magnitude that shear stress exceeds shear strength.

CONCLUSIONS

A new simulator is utilized to examine dominant permeability altering mechanisms as they relate to critical thermal, hydraulic, mechanical, and chemical behaviors in a prototypical geothermal reservoir. Results demonstrate the strong influence of mechanical effects in the short-term, the influence of thermal effects in the intermediate term, and the prolonged and long-term influence of chemical effects. THMC changes are recast in terms of respective characteristic times, where changes in fracture aperture are shown broadly represented by diffusive timescales representing the respective T, H, and C processes. The potential is evident for permeability alteration of several orders of magnitude due to thermo–hydro-mechanical dilation and chemical precipitation/ dissolution.

These two processes compete to dominate reservoir behavior; the former controlling the short-term response, and the long-term response characterized by the latter. For the injection of silica saturated water, amorphous silica is capable of drastically reducing permeability close to the injection well. This influence may be countered by the evolution of mechanical shearing in the near-well region, which may exert a strong influence on reservoir and characteristic behavior. Its inclusion into the modeling structure will be the focus of a future work. As an entry point to such an analysis, we have illustrated potential locations for reservoir shear failure by following the evolution of deviatoric stress and its relationship to thermally-derived shear stress.

We have also taken a first step towards the primary justification for a THMC modeling strategy; that there exists a strong coupling between chemical and mechanical effects that cannot be represented by parallel reactive transport and geomechanics simulations. We provide evidence in support of this assertion, illustrating how mineral behaviors alter fluid flow paths and, in so doing, change the characteristics of thermo–hydro-mechanical aperture changes, and vice versa. We show how each incurs changes in the system that fundamentally alter the evolutionary paths of reaction and chemical/mechanical deformation in a manner that decreases the accuracy of conducting the simulations separately.

ACKNOWLEDGMENTS

This work is the result of partial support from the US Department of Energy under project DOE-DE-FG36-04GO14289. This support is gratefully acknowledged.

REFERENCES

1. Taron J, Elsworth D, Min K-B. Numerical simulation of thermal–hydrologic– mechanical–chemical processes in deformable, fractured porous media. Int J Rock Mech Min Sci 2009; doi:10.1016/j.ijrmms.2009.01.008.
2. Weyl PK. Pressure solution and the force of crystallization—a phenomenological theory. J Geophys Res 1959;64:2001–25.
3. Revil A. Pervasive pressure-solution transfer: a poro–visco-plastic model. Geophys Res Lett 1999;26(2):255–8.
4. Yasuhara H, Elsworth D, Polak A. Evolution of permeability in a natural fracture: significant role of pressure solution. J Geophys Res 2004; 109:B03204.
5. Paterson MS. Nonhydrostatic thermodynamics and its geologic applications. Rev Geophys Space Phys 1973;11:355–89.
6. De Boer RB. On the thermodynamics of pressure solution—interaction between chemical and mechanical forces. Geochem Cosmochem Acta 1977;41:249–56.
7. Xu T, Sonnenthal E, Spycher N, et al. TOUGHREACT—a simulation program for non-isothermal multiphase reactive geochemical transport in variably saturated geologic media: applications to geothermal injectivity and CO2 geological sequestration. Comput Geosci 2006;32:145–65.
8. Rose P, Xu T, Kovac KM, et al. Chemical stimulation in near-wellbore geothermal formations: silica dissolution in the presence of calcite at high temperature and high pH. In: Proceedings of the 32nd workshop on geothermal reservoir engineering, Stanford University, 2007.

9. Nami P, Schellschmidt R, Schindler M, et al. Chemical stimulation operations for reservoir development of the deep crystalline HDR/EGS system at SoultzSouz-Fore^ts (France). In: Proceedings of the 32nd workshop on geothermal reservoir engineering, Stanford University, 2007.
10. Elsworth D, Goodman RE. Characterization of rock fissure hydraulic conductivity using idealized wall roughness profiles. Int J Rock Mech Min Sci 1986;23(3):233–43.
11. Barton N, Bandis S, Bakhtar K. Strength, deformation and conductivity coupling of rock joints. Int J Rock Mech Min Sci 1985;22(3):121–40.
12. Wiltschko DV, Morse JW. Crystallization pressure versus "crack seal" as the mechanism for banded veins. Geology 2001;29(1):79–82.
13. Dewers T, Ortoleva P. Force of crystallization during the growth of siliceous concretions. Geology 1990;18:204–7.
14. Maliva RG, Siever R. Diagenetic replacement controlled by force of crytallization. Geology 1988;16:688–91.
15. Tester J, Blackwell D, Petty S, et al. The future of geothermal energy: impact of enhanced geothermal systems (EGS) on the United States in the 21st century. Cambridge, MA: MIT; 2006. ARTICLE IN PRESS Fig. 8. Plan-view of reservoir with deviatoric stress contours (in MPa) at 5 and 20 years of simulation. White lines are oriented as orthogonal to minimum principal stress direction, and are scaled to the magnitude that shear stress exceeds shear strength. J. Taron, D. Elsworth / International Journal of Rock Mechanics & Mining Sciences 46 (2009) 855–864 863
16. FLAC3D Manual: fast Lagrangian analysis of continua in 3 dimensions—version 2.0. Itasca Consulting Group Inc., Minneapolis, MN, 1997.
17. Settari A. Modeling of fracture and deformation processes in oil sands. In: Proceedings of the fourth UNITAR/UNDP conference on heavy crude and tar sands, paper no. 43, Edmonton, 1988.

18. Settari A, Mourits FM. Coupling of geomechanics and reservoir simulation models. Comput Meth Adv Geomech 1994:2151–8.
19. Minkoff SE, Stone CM, Bryant S, et al. Coupled fluid flow and geomechanical deformation modeling. J Pet Sci Eng 2003;38:37–56.
20. IAPWS, Industrial formulation 1997 for the thermodynamic properties of water and steam. In: International association for the properties of water and steam (IAPWS), 1997.
21. Min K-B, Rutqvist R, Elsworth D. Chemically and mechanically mediated influences on the transport and mechanical characteristics of rock fractures. Int J Rock Mech Min Sci 2008; doi:10.1016/j.ijrmms.2008.04.002.
22. Steefel CI, Lasaga AC. A coupled model for transport of multiple chemical species and kinetic precipitation/dissolution reactions with applications to reactive flow in single phase hydrothermal system. Am J Sci 1994; 294:529–92.
23. Xu T, Pruess K. Modeling multiphase non-isothermal fluid flow and reactive geochemical transport in variably saturated fractured rocks: 1. Methodology. Am J Sci 2001;301:16–33.
24. Helgeson HC, Kirkham DH, Flowers DC. Theoretical prediction of the thermodynamic behavior of aqueous electrolytes at high pressures and temperatures: IV. Calculation of activity coefficients, osmotic coefficients, and apparent molal and standard and relative partial molal properties to 600 1C and 5 kb. Am J Sci 1981;281:1249–516.
25. Xu T, Pruess K. Numerical simulation of injectivity effects of mineral scaling and clay swelling in a fractured geothermal reservoir. Report LBNL-55113, Lawrence Berkeley National Lab, Berkeley; 2004.
26. Rose P. Fracture evolution following a hydraulic stimulation within an EGS reservoir. In: Proceedings of the US DOE EGS program review, 18 July, Golden, CO, USA, 2006.
27. Nemat-Nasser S, Keer LM, Parihar KS. Unstable growth of thermally induced interacting cracks in brittle solids. Int J

Solids Struct 1977;14:409–30.

28. Pruess K. Enhanced geothermal systems (EGS) using CO2 as working fluid—a novel approach for generating renewable energy with simultaneous sequestration of carbon. Geothermics 2006;35(4):351–67.

29. Hunsbedt A, Kruger P, London AL. Recovery of energy from fracturestimulated geothermal reservoirs. J Pet Technol 1977;29:940–6.

30. Elsworth D, Xiang J. A reduced degree of freedom model for thermal permeability enhancement in block rocks. Geothermics 1989;18:691–709.

31. Ba¨chler D. Coupled thermal–hydraulic–chemical modeling at the Soultzsous-Foreˆts HDR reservoir (France). PhD dissertation, ETH, Zurich; 2003.

32. Sheridan JM, Hickman SH. In situ stress, fracture, and fluid flow analysis in well 38C-9: an enhanced geothermal system in the Coso geothermal field. In: Proceedings of the 29th workshop on geothermal reservoir engineering, Stanford University, 2004.

33. Kovac KM, Xu T, Pruess K, et al. Reactive chemical flow modeling applied to injection in the Coso EGS experiment. In: Proceedings of the 31st workshop on geothermal Reservoir engineering, Stanford University, 2006.

34. Mella M, Rose P, Kovac KM, et al. Calcite dissolution in geothermal reservoirs using chelants. In: Transactions on geothermal resource council, San Diego, 2006.

35. McLin KS, Kovac KM, Moore JN, et al. Modeling the geochemical effects of injection at Coso geothermal field, CA; comparison with field observations. In: Proceedings of the 31st workshop on geothermal reservoir engineering, Stanford University, 2006.

36. Wang HF. Theory of linear poroelasticity. Princeton: Princeton University Press; 2000.

Chapter 7

Theory and Application of Numerical Simulation of Chemical Flooding in High Temperature and High Salt Reservoirs

Yirang Yuan, Aijie Cheng, Danping Yang, and Changfeng Li

Institute of Mathematics, Shandong University, Jinan, China

ABSTRACT

Applications, theoretical analysis and numerical methods are introduced for the simulation of mechanical models and principles of the porous flow in high temperature, high salt, complicated geology and large-scale reservoirs in this paper. Considering petroleum geology, geochemistry, computational permeation fluid mechanics and computer technology, we state the models of permeation fluid mechanics and put forward a sequence of implicit upwind difference iteration schemes based on refined fractional steps of the upstream, which can compute the pressures, the saturation and the concentrations of different chemistry components. A type of software applicable in major industries has been completed and carried out in numerical analysis and simulations of oil extraction in Shengli Oil-field, which brings huge economic benefits and social benefits. This software gives many characters: spatial steps are taken as ten meters, the number of nodes is up to hundreds of thousands and simulation time period can be tens of years and the high-order accuracy can be promised in numerical data. Precise analysis is present for simplified models of this type and that provides a tool to solve the international famous problem.

INTRODUCTION

At present an effective method, water-flooding, to keep the pressure of reservoirs is popular in the world, and the recovery efficiency is more outstanding than any other natural exploring forms. It gives more benefits and helps Chinese oil fields keep high quantity production. It continues to be more important and a strategic job to develop the exploiting efficiency of crude oil in the way water-flooding driving.

Much crude oil remains in the reservoir after water-flooding exploiting, which stays underground due to the constraint of capillary force, or doesn't move due to slight influenced region and the fluidity ratio between displacement phase and driven phase. How to develop the displacement efficiency? A popular method considered is that the injected mixture includes chemical addition agents such as polymer, surface active agent and alkali. Polymer can optimize the fluidity of displacement phase, modify the ratio with respect to driven phases, balance the leading edges well, weaken the inner porous layer, and increase the efficiency of displacement and the pressure gradient. Surface active agent and alkali can decrease interfacial tensions of different phases, then make the bounded oil move and gather. Some hypotheses should be made for the mathematical models. Local thermodynamic equilibrium holds in the reservoir, solid phase has no motion, and the rock and mixture fluid are slight- compressible, of Fick dispersion, ideal and suitable for Darcy Law.

The equilibrium equation of multi-phase, multi-components and slight compressible mixture is formulated by a nonlinear coupled system of partial differential equations. It is hard to solve this system because many modern numerical methods such as mixed element, finite element, finite difference and numerical algebra, will be involved in the simulation. In general speaking, based on physical means the pressure function is solved by an implicit scheme and the concentration values are obtained by an explicit solver or an implicit solver. The scholars try to find good ways analyzing the data and numerical results and doing some research work in simulation, which describe the whole process of chemistry displacements very well and help the engineers control the rules and process of displacement and forecast the recovery efficiency of natural oil and compute the oil percentage of output liquid and the percent of polymer and surface active agent. By numerical research the curves describing different components motion are shown, and some plans are made about the beginning and end of injected liquid and some related parameters of natural oil efficiency are derived. These conclusions, important techniques in chemistry displacements, can be used in forecasting the characters of fields, choosing different optimization plans, establishing the models of chemical displacements of reservoir, completing computational software and carrying out the numerical simulation. Petroleum engineers and mathematicians pay more attention to modern new techniques of exploiting natural oil.

Yuan visited United States and accomplished some work cooperate with Prof. R. E. Ewing during 1985 to 1988, and kept a series of research in theoretical analysis and applications of numerical simulation. With the leading of Yuan several research groups undertook some important projects from 1991 to 1995 such as "Eighth- Five" national key science and technology program (the Program for Tackling Key Programs) (85-203-01-087) entitled "research and application of the polymer displacement software" [1] -[6] . The software was applied in designing plan and research work of polymer displacements in industrial production region of Daqing Oilfield. Many conclusions from actual numerical results are illustrated such as effects of fragments, fragments setting of rinsing protection, quantity of polymer, and used in actual simulations which give rise to outstanding economic and social benefits [7] - [9] . Later the authors undertook a key tackling program of oil administration of Daqing Oilfield (DQYJ-1201002-2006-JS-9565)—solving development of mathematical models and completing explain of reservoir [10] . This software system is also applied in three compound combination flooding of Gudong Little Well experimental region of Shengli Oilfield, polymer flooding of Gudong Middle One experimental region, optimization of combination flooding expanded experimental region of Gudong West region and feasibility of active water flooding of Gudong eighth region, and many interesting results are obtained [11] .

Theory, method and application of numerical simulation are studied for high temperature, high salt, complicated geology and large-scale reservoirs and the principle of chemical flooding in this paper. Based on the former research, the conclusions and more discussion of the national major special project on science and technology (2008ZX05011-004) "Study on key technology of chemical flooding numerical simulation in high temperature and high salt reservoirs (on numerical simulation)" are given, which consists of permeation fluid mechanical models of numerical simulation of high temperature and high salt polymer flooding and compound combination flooding, numerical methods, applicable software, theoretical analysis and applications in oilfields.

MATHEMATICAL MODEL OF POLYMER FLOODING AND COMPOUND COMBINATION FLOODING

This section includes five subsections: Section 2.1, basic hypothesis, Section 2.2, conservation equation of matter, Section 2.3, the pressure equation, Section 2.4, the concentration equation, Section 2.5, the concentration equation of chemical components.

Basic Hypothesis

Mathematical model of polymer flooding and compound combination flooding is derived under the following hypothesis. Local thermodynamic equilibrium holds in the reservoir, solid phase doesn't motion, rock and mixture fluid are slight-compressible, of Fick dispersion, ideal and suitable for Darcy Law [1] -[3] [7] -[10] .

Conservation Equation of Matter

Under the primary hypothesis, a conservation equation of the *i*-th component is stated as follows dependent of the i-th concentration $\tilde{C}_i$:

$$\frac{\partial}{\partial t}\left(\phi \tilde{C}_i \rho_i\right) + \mathrm{div} \cdot \left[\sum_{l=1}^{n_p} \rho_i \left(C_{il} \boldsymbol{u}_l - \tilde{\boldsymbol{D}}_{il}\right)\right] = Q_i, \tag{1}$$

where C_{il} represents the concentration the concentration of the *i*-th component in the l-phase, Q_i means source sink term, n_p means the number of phases, and subscripts l denotes the order of phases. The symbol $\tilde{C}_i$ represents the total concentration of the i-th component, *i.e.* the summation of concentrations of the i-th component in different phases (including adsorbed phase):

$$\tilde{C}_i = \left(1 - \sum_{k=1}^{n_{cv}} \hat{C}_k\right) \sum_{l=1}^{n_p} S_l C_{il} + \hat{C}_i, \qquad i = 1, \cdots, n_{cv}, \tag{2}$$

Where n_{cv}, the number of component whose volume couldn't be ignored, and $\hat{C}_i$ denotes the adsorption concentration of the component k.

The density of the i-th component is dependent of the pressure r_i under slight compressible case :

$$\rho_i = \rho_i^o \left[1 + C_i^o \left(p - p_r\right)\right], \tag{3}$$

where ρ_i^o is the density of the i-th component under considering reference pressure P_r, and P means the pressure and C_i^o is the coefficient of compressibility of the i-th component.

Suppose that the rock is compressible, then the function of the porosity ϕ and the pressure is

$$\phi = \phi^o \left[1 + C^r \left(p - p^r\right)\right], \tag{4}$$

where C_r means the coefficient of compressibility of the rock.

Darcy velocity, u_l, is described by Darcy Law,

$$\boldsymbol{u}_l = \frac{KK_{rl}}{\mu_l}\left(\nabla p_l - \gamma_l \nabla D\right), \tag{5}$$

where P_l means the pressure of phases, K is the permeability tensor, D is the depth, K_{rl} is the relative permeability, $\mathbf{m}_l$ is the viscosity and $\mathbf{g}_l$ is the proportion.

The dispersion flux is expressed in the following Fick formation :

$$\tilde{\boldsymbol{D}}_{il} = \phi S_l \begin{pmatrix} F_{xx,il} & F_{xy,il} & F_{xz,il} \\ F_{yx,il} & F_{yy,il} & F_{yz,il} \\ F_{zx,il} & F_{zy,il} & F_{zz,il} \end{pmatrix} \begin{pmatrix} \partial C_{il}/\partial x \\ \partial C_{il}/\partial y \\ \partial C_{il}/\partial z \end{pmatrix}. \tag{6}$$

The dispersion tensor F_{il}, including molecular diffusion (D_{kl}), can be formulated as

$$F_{mn,il} = \frac{D_{il}}{\tau}\delta_{mn} + \frac{\alpha T_l}{\phi S_l}\left|\boldsymbol{u}_l\right|\delta_{mn} + \frac{(\alpha L_l - \alpha T_l)}{\phi S_l}\frac{u_{lm}u_{ln}}{\boldsymbol{u}_l}, \tag{7}$$

where α_{Ll} and $\mathbf{a}_{Tl}$ denote the longitudinal and the lateral dispersion parameters of the l-th phase, τ is the tortuosity, u_{lm} and u_{ln} are the components, and δ_{mn} means Kronecher Delta. The net flow of each phase is:

$$\left|\boldsymbol{u}_l\right| = \sqrt{u_{xl}^2 + u_{yl}^2 + u_{zl}^2}. \tag{8}$$

The Pressure Equation

Considering all the conservative equations of each matter with positive volume together, using Darcy Law and capillary pressure to express the flux and the pressure relations of different phases respectively, and combining with the following constraints

$$\sum_{i=1}^{n_{cv}} C_{il} = 1, \tag{9}$$

We can get the pressure equation of referenced phase:

$$\phi C_t \frac{\partial p_w}{\partial t} + \operatorname{div}\cdot\left(K\lambda_T \nabla p_w\right) = -\operatorname{div}\cdot\left(\sum_{l=1}^{n_p} K\lambda_l \nabla h\right) + \operatorname{div}\cdot\left(\sum_{l=1}^{n_p} K\lambda_l \nabla p_{clw}\right) + \sum_{i=1}^{n_{cv}} Q_i. \tag{10}$$

where

$$\lambda_l = \frac{K_{rl}}{\mu_l}\sum_{i=1}^{n_{cv}} \rho_i C_{il}, \tag{11}$$

and the total relative fluidity l_T is

$$\lambda_T = \sum_{i=1}^{n_{cv}} \lambda_l. \tag{12}$$

The total coefficient of compressibility C_t, is a function dependent of the compressibility of rocks C_r and components of the mixture C_i^o:

$$C_t = C_r + \sum_{i=1}^{n_{cr}} C_i^o \tilde{C}_i.$$

The Concentration Equation

Let S_w and S_o be concentrations of water phase and oil phase denoted by subscripts w and o denote and the relation $S_w+S_o=1$ holds obviously. The equations on concentrations of water and oil phases are expressed as follows by conservation of the mass (1)

$$\frac{\partial}{\partial t}(\phi S_o \rho_o) + \mathrm{div}\cdot(\rho_o \boldsymbol{u}_o) = Q_o, \tag{13}$$

$$\frac{\partial}{\partial t}(\phi S_w \rho_w) + \mathrm{div}\cdot(\rho_w \boldsymbol{u}_w) = Q_w. \tag{14}$$

By Darcy Law the velocities of different phases are derived by

$$\boldsymbol{u}_w = -K\lambda_{rw}(\nabla P_1 - \gamma_1 \nabla D), \tag{15}$$

$$\boldsymbol{u}_o = -K\lambda_{ro}(\nabla P_2 - \gamma_2 \nabla D) = -K\lambda_{ro}(\nabla P_1 + \nabla P_c - \gamma_2 \nabla D), \tag{16}$$

where $\lambda_{rw} = K_{rw}(S_w)/\mu_w(S_w, C_{1w}, \cdots, C_{Lw})$, $\lambda_{ro} = K_{ro}(S_w)/\mu_o(S_w)$ are the fluidities of two phases, and K, K_{rw} and K_{ro} are the absolute permeability tensor, relative permeabilities. K is the absolute permeability of medium, and μ_w, μ_o are the viscosities of water-oil phases dependent on their concentrations and the saturations of polymer and the two opposing principles in nature. P_1, P_2 and γ_1, γ_2 respectively denote the pressures and the densities of water phase and oil phase and D is the depth function,

$$D = D(x, y, z) = z.$$

The Concentrations of Chemical Substance Components

Note that all the components (polymer and different principles) are mixed in the water phase and no transmission takes place, then

$$C_{il} = \begin{cases} C_{iw}, & l = w \\ 0, & l \neq w. \end{cases} \tag{17}$$

Substitute (17) into (1),

$$\frac{\partial}{\partial t}(\phi \rho_i \lambda_i C_{iw}) + \text{div} \cdot \left[\rho_i \left(C_{iw} \boldsymbol{u}_w - \tilde{\boldsymbol{D}}_{iw} \right) \right] = Q_i, \tag{18}$$

where $\lambda_i = (S_w) + A(C_{iw})$ and $A(C_{iw})$ is dependent of the absorptions.

NUMERICAL METHODS

This section consists of three subsections: Section 3.1 solving the pressure equation, Section 3.2 solving the concentration equation, and Section 3.3 structuring a numerical algorithm of components.

Solving the Pressure

Let the parameters with subscripts w, o be related with the water and the oil respectively, such as S_w, S_o and P_w, P_o denote the saturations and the pressures of water and oil phases. Note that the mixture fluid is only made by oil and water two phases in the model of polymer flooding and compound combination flooding and we can describe the pressure in a simple formula

$$\phi c_t \frac{\partial P}{\partial t} + \text{div} \cdot \left(\boldsymbol{u}_w + \boldsymbol{u}_o \right) = Q_w + Q_o. \tag{19}$$

Using Darcy Law and the formula of capillary force, we rewrite the above equation as follows with respect to an unknown variable $P(P_o)$,

$$\phi c_t \frac{\partial P}{\partial t} - \mathrm{div} \cdot \left[\left(\lambda_{rw} + \lambda_{ro} \right) \nabla P \right] = Q_w + Q_o + \mathrm{div} \cdot \left[\lambda_{ro} \nabla P_c - \left(\gamma_w + \gamma_o \right) \nabla D \right] \quad (20)$$

The initial values of the saturations are known at the beginning of simulations while the pressure values should be initialized in the following process. When the pressure P^n at the n-th time step are known, then the flow velocities are obtained and the values of the saturation and components at the next step $t=t^{n+1}$ are computed. The pressure, denoted by a parabolic equation, is obtained by a seven-point central difference method. Considering the physical features of two-phase we assign λ_{rw} and λ_{ro}, the values of the left-side term, in accordance with upstream principles. At the injected wells and produced wells with fixed quantities, the right source term can be assigned directly and the values of quantity are determined by the difference between the pressure of local regions and the pressure of bottom holes at injected wells and produced wells with fixed pressures. The production quantities of different phases are distributed by the relative fluidity of oil-water phases. In addition, the pressure equation is degraded into an equation of elliptic type, and the matrix is not strictly diagonal-dominated under an impressible assumption (the coefficients of compressibility are assumed to be zero). In the way of taking the diagonal unit be 1 and non-diagonal units be zero, the equation of feature edges is consistent with the equation of normal oil deposits. The data in feature edges don't need to be replaced by the values of solutions, which makes the quantity of physical data as a constant in the computation. If the program (necessary for the design) runs at feature edges virtual data will be used in the whole computation.

Given P^n, using upstream seven-point central difference algorithm to compute P^{n+1},

$$
\begin{aligned}
&\phi_{ijk} c_{t,ijk} \frac{P_{ijk}^{n+1} - P_{ijk}^{n}}{\Delta t} - \frac{\left(\lambda_{rw,i_+jk} + \lambda_{ro,i_+jk}\right)\left(P_{i+1,jk}^{n+1} - P_{ijk}^{n+1}\right) - \left(\lambda_{rw,i_-jk} + \lambda_{ro,i_-jk}\right)\left(P_{ijk}^{n+1} - P_{i-1,jk}^{n+1}\right)}{(\Delta x)^2} \\
&- \frac{\left(\lambda_{rw,ij_+k} + \lambda_{ro,ij_+k}\right)\left(P_{i,j+1,k}^{n+1} - P_{ijk}^{n+1}\right) - \left(\lambda_{rw,ij_-k} + \lambda_{ro,ij_-k}\right)\left(P_{ijk}^{n+1} - P_{i,j-1,k}^{n+1}\right)}{(\Delta y)^2} \\
&- \frac{\left(\lambda_{rw,ijk_+} + \lambda_{ro,ijk_+}\right)\left(P_{ij,k+1}^{n+1} - P_{ijk}^{n+1}\right) - \left(\lambda_{rw,ijk_-} + \lambda_{ro,ijk_-}\right)\left(P_{ijk}^{n+1} - P_{ij,k-1}^{n+1}\right)}{(\Delta z)^2} \\
&= Q_{w,ijk}^{n+1} + Q_{o,ijk}^{n+1} - \frac{\lambda_{ro,i_+jk}\left(P_{c,i+1,jk}^{n+1} - P_{c,ijk}^{n+1}\right) - \lambda_{ro,i_-jk}\left(P_{c,ijk}^{n+1} - P_{c,i-1,jk}^{n+1}\right)}{(\Delta x)^2} \\
&- \frac{\lambda_{ro,ij_+k}\left(P_{c,i,j+1,k}^{n+1} - P_{c,ijk}^{n+1}\right) - \lambda_{ro,ij_-k}\left(P_{c,ijk}^{n+1} - P_{c,i,j-1,k}^{n+1}\right)}{(\Delta y)^2} - \frac{\lambda_{ro,ijk_+}\left(P_{c,ij,k+1}^{n+1} - P_{c,ijk}^{n+1}\right) - \lambda_{ro,ijk_-}\left(P_{c,ijk}^{n+1} - P_{c,ij,k-1}^{n+1}\right)}{(\Delta z)^2} \\
&+ \frac{\left(\gamma_{w,ij,k+1} + \gamma_{w,ijk}\right)\left(D_{ij,k+1}^{n+1} - D_{ijk}^{n+1}\right) - \left(\gamma_{w,ijk} + \gamma_{w,ij,k-1}\right)\left(D_{ijk}^{n+1} - D_{ij,k-1}^{n+1}\right)}{2(\Delta z)^2}.
\end{aligned}
\tag{21}
$$

The subscript denotes the upstream position in the first direction between the i-th point and the (i+1)-th point. An assumption is given that there is no flow moving through the boundary. That is to say that its boundary condition is homogeneous of Neumann type. The quantities at the injected wells $Q_w^{n+1} = 0$, Q_w^{n+1} are known and those at the produced wells $Q_o^{n+1} = Q_o^{n+1}(P_f, P_o, S_o)$, $Q_w^{n+1} = Q_w^{n+1}(P_f, P_w, S_w)$ are given implicitly by the flowing bottom hole pressure P_f, the pressures of phases and the relative mobility ratio. Distributed quantities are computed by an allocation program after the pressure values are obtained. It is easy to solve the saturation equation when the values of source and sink terms.

Solving the Saturation

Using upstream order, implicit upwind Newton iteration to solve the saturation of water phase, then the values of oil saturation S_o=1-S_w are computed. The saturation of water phase is described by

$$
\phi \frac{\partial S_w}{\partial t} + \operatorname{div} \cdot \boldsymbol{u}_w = Q_w,
$$

or expressed in another form combined with Darcy Law

$$
\phi \frac{\partial S_w}{\partial t} + \operatorname{div} \cdot \left(\lambda_w(S_w)\left(\nabla P_w - \gamma_w \nabla D\right)\right) = Q_w. \tag{22}
$$

The pressure of water phase P_w at t^{n+1} and the quantity Q_w are known, then the saturation of water S_w at t^{n+1} are computed by the following discrete algorithm,

$$\phi_{ijk}\frac{S_{w,ijk}^{n+1}-S_{w,ijk}^{n}}{\Delta t}-\frac{\lambda_w\left(S_{w,i_+}\right)\left(P_{i+1,jk}^{n+1}-P_{ijk}^{n+1}\right)-\lambda_w\left(S_{w,i_-}\right)\left(P_{ijk}^{n+1}-P_{i-1,jk}^{n+1}\right)}{(\Delta x)^2}$$
$$-\frac{\lambda_w\left(S_{w,j_+}\right)\left(P_{i,j+1,k}^{n+1}-P_{ijk}^{n+1}\right)-\lambda_w\left(S_{w,j_-}\right)\left(P_{ijk}^{n+1}-P_{i,j-1,k}^{n+1}\right)}{(\Delta y)^2}$$
$$-\frac{\lambda_w\left(S_{w,k_+}\right)\left(P_{ij,k+1}^{n+1}-P_{ijk}^{n+1}\right)-\lambda_w\left(S_{w,k_-}\right)\left(P_{ijk}^{n+1}-P_{ij,k-1}^{n+1}\right)}{(\Delta z)^2}$$
$$=Q_{w,ijk}^{n+1}+\frac{\lambda_w\left(S_{w,k_+}\right)\left(D_{ij,k+1}^{n+1}-D_{ijk}^{n+1}\right)-\lambda_w\left(S_{w,k_-}\right)\left(D_{ijk}^{n+1}-D_{ij,k-1}^{n+1}\right)}{(\Delta z)^2}, \quad (23)$$

where the subscript i_+ denotes the upstream position according to the moving trend between x_i and x_{i+1}, *i.e.* it is either i or i+1. It is hard to solve the nonlinear equations directly. $S_w^{n+1,l+1}$ at the (l+1) iteration step is obtained by Newton iteration under taking the value at the previous step as the initial condition $S_w^{n+1,0}=S_w^n$. Note the following Taylor expansion,

$$\lambda_w\left(S_{w,i_+jk}^{n+1,l+1}\right)=\lambda_w\left(S_{w,i_+jk}^{n+1,l}\right)+\lambda_w'\left(S_{w,i_+jk}^{n+1,l}\right)\left(S_{w,i_+jk}^{n+1,l+1}-S_{w,i_+jk}^{n+1,l}\right)+O\left(\left(S_{w,i_+jk}^{n+1,l+1}-S_{w,i_+jk}^{n+1,l}\right)^2\right),$$

$$\lambda_w\left(S_{w,i_-jk}^{n+1,l+1}\right)=\lambda_w\left(S_{w,i_-jk}^{n+1,l}\right)+\lambda_w'\left(S_{w,i_-jk}^{n+1,l}\right)\left(S_{w,i_-jk}^{n+1,l+1}-S_{w,i_-jk}^{n+1,l}\right)+O\left(\left(S_{w,i_-jk}^{n+1,l+1}-S_{w,i_-jk}^{n+1,l}\right)^2\right),$$

$$\lambda_w\left(S_{w,ij_+k}^{n+1,l+1}\right)=\lambda_w\left(S_{w,ij_+k}^{n+1,l}\right)+\lambda_w'\left(S_{w,ij_+k}^{n+1,l}\right)\left(S_{w,ij_+k}^{n+1,l+1}-S_{w,ij_+k}^{n+1,l}\right)+O\left(\left(S_{w,ij_+k}^{n+1,l+1}-S_{w,ij_+k}^{n+1,l}\right)^2\right),$$

$$\lambda_w\left(S_{w,ij_-k}^{n+1,l+1}\right)=\lambda_w\left(S_{w,ij_-k}^{n+1,l}\right)+\lambda_w'\left(S_{w,ij_-k}^{n+1,l}\right)\left(S_{w,ij_-k}^{n+1,l+1}-S_{w,ij_-k}^{n+1,l}\right)+O\left(\left(S_{w,ij_-k}^{n+1,l+1}-S_{w,ij_-k}^{n+1,l}\right)^2\right),$$

$$\lambda_w\left(S_{w,ijk_+}^{n+1,l+1}\right)=\lambda_w\left(S_{w,ijk_+}^{n+1,l}\right)+\lambda_w'\left(S_{w,ijk_+}^{n+1,l}\right)\left(S_{w,ijk_+}^{n+1,l+1}-S_{w,ijk_+}^{n+1,l}\right)+O\left(\left(S_{w,ijk_+}^{n+1,l+1}-S_{w,ijk_+}^{n+1,l}\right)^2\right),$$

$$\lambda_w\left(S_{w,ijk_-}^{n+1,l+1}\right)=\lambda_w\left(S_{w,ijk_-}^{n+1,l}\right)+\lambda_w'\left(S_{w,ijk_-}^{n+1,l}\right)\left(S_{w,ijk_-}^{n+1,l+1}-S_{w,ijk_-}^{n+1,l}\right)+O\left(\left(S_{w,ijk_-}^{n+1,l+1}-S_{w,ijk_-}^{n+1,l}\right)^2\right).$$

Delete the remainder truncation, substitute the approximation expressions into the difference algorithm, take $S_{w,ijk}^{n+1,0}=S_{w,ijk}^{n}$ as iteration initial conditions and obtain the iteration values $S_{w,ijk}^{n+1,l+1}$ at the l+1 -th step by solving the following linear equations

$$\begin{aligned}
&\phi_{ijk}\frac{S_{w,ijk}^{n+1}-S_{w,ijk}^{n}}{\Delta t}\\
&-\left\{\left[\lambda_w\left(S_{w,i_+jk}^{n+1,l}\right)+\lambda_w'\left(S_{w,i_+jk}^{n+1,l}\right)\left(S_{w,i_+jk}^{n+1,l+1}-S_{w,i_+jk}^{n+1,l}\right)\right]\left(P_{i+1,jk}^{n+1}-P_{ijk}^{n+1}\right)-\left[\lambda_w\left(S_{w,i_-jk}^{n+1,l}\right)+\lambda_w'\left(S_{w,i_-jk}^{n+1,l}\right)\left(S_{w,i_-jk}^{n+1,l+1}-S_{w,i_-jk}^{n+1,l}\right)\right]\left(P_{ijk}^{n+1}-P_{i-1,jk}^{n+1}\right)\right\}/(\Delta x)^2\\
&-\left\{\left[\lambda_w\left(S_{w,ij_+k}^{n+1,l}\right)+\lambda_w'\left(S_{w,ij_+k}^{n+1,l}\right)\left(S_{w,ij_+k}^{n+1,l+1}-S_{w,ij_+k}^{n+1,l}\right)\right]\left(P_{i,j+1,k}^{n+1}-P_{ijk}^{n+1}\right)-\left[\lambda_w\left(S_{w,ij_-k}^{n+1,l}\right)+\lambda_w'\left(S_{w,ij_-k}^{n+1,l}\right)\left(S_{w,ij_-k}^{n+1,l+1}-S_{w,ij_-k}^{n+1,l}\right)\right]\left(P_{ijk}^{n+1}-P_{i,j-1,k}^{n+1}\right)\right\}/(\Delta y)^2\\
&-\left\{\left[\lambda_w\left(S_{w,ijk_+}^{n+1,l}\right)+\lambda_w'\left(S_{w,ijk_+}^{n+1,l}\right)\left(S_{w,ijk_+}^{n+1,l+1}-S_{w,ijk_+}^{n+1,l}\right)\right]\left(P_{ij,k+1}^{n+1}-P_{ijk}^{n+1}\right)-\left[\lambda_w\left(S_{w,ijk_-}^{n+1,l}\right)+\lambda_w'\left(S_{w,ijk_-}^{n+1,l}\right)\left(S_{w,ijk_-}^{n+1,l+1}-S_{w,ijk_-}^{n+1,l}\right)\right]\left(P_{ijk}^{n+1}-P_{ij,k-1}^{n+1}\right)\right\}/(\Delta z)^2\\
&=Q_{w,ijk}^{n+1}\\
&+\left\{\left[\lambda_w\left(S_{w,ijk_+}^{n+1,l}\right)+\lambda_w'\left(S_{w,ijk_+}^{n+1,l}\right)\left(S_{w,ijk_+}^{n+1,l+1}-S_{w,ijk_+}^{n+1,l}\right)\right]\left(D_{ij,k+1}^{n+1}-D_{ijk}^{n+1}\right)-\left[\lambda_w\left(S_{w,ijk_-}^{n+1,l}\right)+\lambda_w'\left(S_{w,ijk_-}^{n+1,l}\right)\left(S_{w,ijk_-}^{n+1,l+1}-S_{w,ijk_-}^{n+1,l}\right)\right]\left(D_{ijk}^{n+1}-D_{ij,k-1}^{n+1}\right)\right\}/(\Delta z)^2,\\
&l=0,1,\cdots,L.
\end{aligned}\tag{24}$$

The program runs based on upstream sequence rule, and the iteration value at the (l+1)-th step is computed until the relative error meets a designed requirement or the iteration reaches the steps when the values at upstream points are known. The final iterative value is denoted by S_w^{n+1}.

Numerical Algorithm of Concentration Components

The components of water phase keep conservation of the mass of anions, cations and molecules and other particles, whose equation is of diffusion-convection and convection dominated. It has more strengths such as high order of accuracy and high efficiency of simulation applying decomposition of operators into the nonlinear system and solving two subproblems: a hyperbolic equation of convection type and a diffusion equation. The former is solved implicitly by an upwind method, which can be carried out explicitly by an upstream technique. The latter is solved by alternating directions finite difference method, which can improve the computational speed. The concentration equation of k-component, a typical convection-diffusion equation, is simplified as follows

$$\phi\frac{\partial}{\partial t}\left(S_w C_k\right)+\mathrm{div}\cdot\left(C_k \boldsymbol{u}_w-\phi S_w K\nabla C_k\right)=Q_k,\tag{25}$$

where the dispersion tensor is a diagonal form. Given the saturation S_w and the flow field u_w, the concentration C_k is to be computed by using an implicit upwind method to solve a convection problem, where the subscript k is ignored and C denotes a component concentration.

$$\phi_{ijk}\frac{S_w^{n+1}C_{ijk}^{n+1,0}-S_w^nC_{ijk}^{n+1}}{\Delta t}+\left[C_{i_+jk}^{n+1,0}u_{w,ijk}^{n+1}-C_{i_-jk}^{n+1,0}u_{w,i-1,jk}^{n+1}\right]\Big/\Delta x+\left[C_{ij_+k}^{n+1,0}u_{w,ijk}^{n+1}-C_{ij_-k}^{n+1,0}u_{w,i,j-1,k}^{n+1}\right]\Big/\Delta y$$
$$+\left[C_{ijk_+}^{n+1,0}u_{w,ijk}^{n+1}-C_{ijk_-}^{n+1,0}u_{w,ij,k-1}^{n+1}\right]\Big/\Delta z=Q_{ijk}^{n+1}, \tag{26}$$

The values $C^{n+1,0}$ is obtained, then the diffusion equation is discretized alternatively in three directions. In x-direction,

$$\phi_{ijk}\frac{S_w^{n+1}C_{ijk}^{n+1,1}-S_w^nC_{ijk}^{n+1,0}}{\Delta t}$$
$$-\left[\phi_{i+1/2,jk}S_{w,i_+jk}^{n+1}K_{xx,i+1/2,jk}\left(C_{i+1,jk}^{n+1,1}-C_{ijk}^{n+1,1}\right)\right]-\phi_{i-1/2,jk}S_{w,i_-jk}^{n+1}K_{xx,i-1/2,jk}\left(C_{ijk}^{n+1,1}-C_{i-1,jk}^{n+1,1}\right)\Big/(\Delta x)^2=0, \tag{27a}$$

Then in y-direction,

$$\phi_{ijk}\frac{S_w^{n+1}C_{ijk}^{n+1,2}-S_w^nC_{ijk}^{n+1,1}}{\Delta t}$$
$$-\left[\phi_{i,j+1/2,k}S_{w,ij_+k}^{n+1}K_{yy,i,j+1/2,k}\left(C_{i,j+1,k}^{n+1,2}-C_{ijk}^{n+1,2}\right)\right]-\phi_{i,j-1/2,k}S_{w,ij_-k}^{n+1}K_{yy,i,j-1/2,k}\left(C_{ijk}^{n+1,2}-C_{i,j-1,k}^{n+1,2}\right)\Big/(\Delta y)^2=0, \tag{27b}$$

At last in z-direction, C_{n+1} is obtained,

$$\phi_{ijk}\frac{S_w^{n+1}C_{ijk}^{n+1}-S_w^nC_{ijk}^{n+1,2}}{\Delta t}$$
$$-\left[\phi_{ij,k+1/2}S_{w,ijk_+}^{n+1}K_{zz,ij,k+1/2}\left(C_{ij,k+1}^{n+1}-C_{ijk}^{n+1}\right)\right]-\phi_{ij,k-1/2}S_{w,ijk_-}^{n+1}K_{zz,ij,k-1/2}\left(C_{ijk}^{n+1}-C_{ij,k-1}^{n+1}\right)\Big/(\Delta z)^2=0, \tag{27c}$$

Then the discrete solutions P_w^{n+1}, P_o^{n+1}, S_w^{n+1}, S_o^{n+1}, C_k^{n+1} are obtained and the computation runs in the next step.

Viscosity Computation of High Temperature and High Salt Reservoir

The polymer hydrolysis can decrease the viscosity in the high temperature and high salt reservoirs, and the viscosity is different under different polymers. The viscosity is computed by

$$\mu_{ap} = \mu_w + (\mu_o - \mu_w)\left[1 + (\gamma/\gamma_{1/2})^{n(p)-1}\right]^{-1}, \tag{28}$$

where the values of parameters are defined as follows,

$\gamma_{1/2} = 375.3\mu_o^{-1.378} + 0.0356$, $\quad n(p) = 1.163\mu_o^{0.0311}$,

$\mu_o = \mu_w + 0.634C_p[\mu] + 0.193(C_p[\mu])^2 + 0.921(C_p[\mu])^3$,

$[\mu] = 4.665\times10^{-3}\overline{M}^{0.7969} + 4.219\times10^{-2}(1-h_p^{1/2})h_p\overline{M}^{0.76}/\sqrt{C_s}$.

$\overline{M}$ denotes the average molecular weight of polymer, and h_p denotes the degree of hydrolysis of the polymer with decimal unit. C_s means the degree of mineralization and C_p means the concentration of polymer, whose units are mol/L and mg/L, respectively.

COMPUTATION PROGRAM ILLUSTRATION

This section illustrates the computation program by Figure 1.

ACTUAL EXPERIMENTAL TESTS OF OIL FIELDS

The adaptation efficiency of the software SLCHEM is tested in view of three aspects: dependability, universality and special applicability for large-scale oilfields. The numerical results are dependable by comparing with the actual results and popular business software computations. The software is applied successfully in different fields of Shengli Oilfield and the universality is tested. The software is used in large-scale oilfields and the special applicability is tested.

Experiments of the Polymer Flooding in Small-Scale Oilfields

The rectangle computational domain (Tuo Block 28) is partitioned into $22 \times 24 \times 6$ subdomains with uniform steps and the spacial steps in *x*-direction and *y*-direction are taken as 45.61 m and 47.35 m, respectively. The irreducible water saturation is 0.175, residual oil saturation is 0.25, and formation reserve is 9.9843×10^7 m^3. There are fifty four wells in this block. The simulation works about 22,402 days and is considered in three periods. It works in November, 1966 and water is injected in March, 1967. The polymer, whose concentration is 1500 ppm, is injected from June, 2008 to April, 2015. Water is injected again from April, 2015 to January, 2030. The simulation results are illustrated in Table 1.

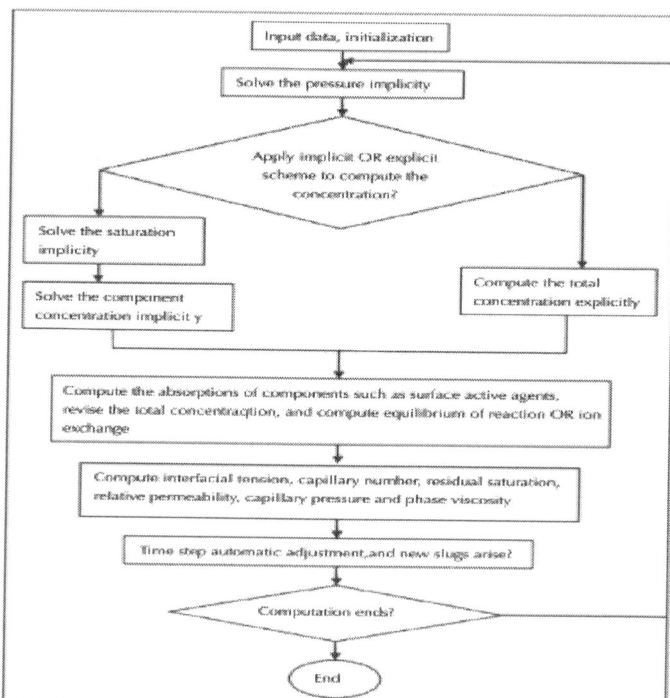

Figure 1: Computation program illustration.

The comparison of moisture content of produced oil of SLCHEM, VIP, and actual results are shown in Figure 2.

The time cost of SLCHEM is about 0.44 hour, and the material balance error is satisfactory. The relative total error comparison with actual moisture content is about 7.8% before the polymer flooding is injected. All the results show that the computation of SLCHEM runs fast, the numerical results are reliable and this scheme can be applied into present oilfields production.

Experiments of Surfactant-Polymer Flooding Agents in Middle-Scale Oilfields

The rectangle computational domain (Sheng Block 2) is partitioned into 82 × 74 × 7 subdomains with uniform steps and the spacial steps in x-direction and y-direction are taken as 33.60 m and 29.29 m, respectively. Formation reserve is 3.4938×10^8 m^3. There are sixty three wells in this block. The simulation works about$2362 days and is considered in three periods. It works on April 1, 1966 and water is injected on January 1, 1974. The polymer, whose concentration is 1600 ppm, is injected from August, 2010 to September, 2016. Surfactant, whose concentration is 0.4%, is injected from August, 2011 to October, 2011. The largest time step is less than five days and the simulation results are illustrated in Table 2.

Table 1: Numerical results of polymer flooding in small-scale oilfield

Scheme	Time cost	Computational steps	Time cost of the pressure	Balance error (two phases)	Balance error (components)
New algorithm	1600.578	8516	517.156	10^{-6}	10^{-13}

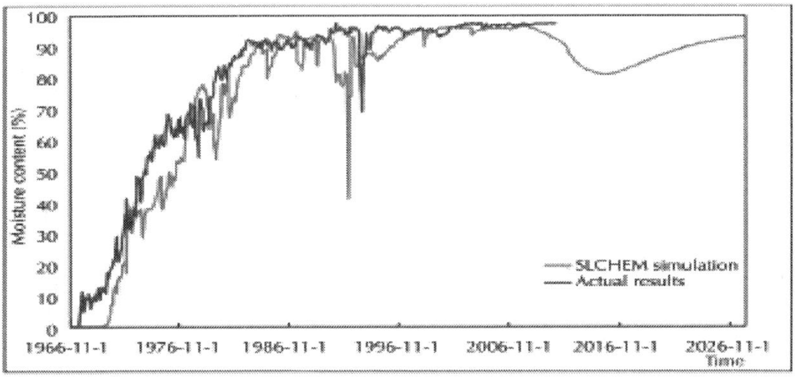

Figure 2: Curve of moisture content simulation of produced oil of polymer flooding in small- scale oilfields.

The comparison of moisture content of produced oil of SLCHEM, ECLIPSE, and actual results are shown in Figure 3. The time cost of SLCHEM is about 4.2 hours, and the material balance error is satisfactory. The relative total error comparison with actual moisture content is about 7.10%. All the results show that the computation of SLCHEM runs fast, the numerical results are reliable and this scheme can be applied into present oilfields production.

Experiments of Large-scale oilfields

The rectangle computational domain (Gudaoguan Block 3) is partitioned into 72 × 62 × 26 subdomains with uniform steps and the spacial steps in x-direction and y-direction are taken as 25 m. Formation reserve is 22,708,200.0 m^3. There are one hundred and ninety one wells in this block. The simulation works about 12,965 days and is considered in three periods. It works on September 1, 1971 and water is injected on September 1, 1974. The polymer, whose concentration is 1000 ppm, is injected from March 1, 1994 to November 30, 2003. Water is injected again from December 1, 2003 to March 1, 2007. The largest time step is less than ten days and the simulation results are illustrated in Table 3. The comparison of moisture content of produced oil of SLCHEM and actual results are shown inFigure 4. The time cost of SLCHEM is about 8.45 hours, and the material balance error is 10^{-6}. The relative total error comparison with actual moisture content is about 7.0%. All the results show that the computation of SLCHEM runs fast, the numerical results are reliable and this scheme can be applied into present oilfields production.

Numerical Simulation of Surfactant Polymer Flooding

The polymer, whose concentration is 1000 ppm, is injected from March 1, 1994 to November 30, 2003. Surfactant, whose concentration is 0.5%, is injected from May 1, 1996 to December 1, 2003. Then water is injected again from December 1, 2003 to March 1, 2007. The largest time step is less than ten days and the simulation results are illustrated in Table 4.

Numerical results of SLCHEM of water flooding, polymer flooding and surfactant-polymer flooding are compared in Figure 5. The time cost of SLCHEM is about 13.5 hours, and the material balance error is satisfactory. The software has three flooding functions and physical-chemical parameters can be processed effectively. The software simulates reliably and generally based on the above experiments of different scales oilfields and the simulation scale is up to hundreds of thousands nodes.

Table 2: Numerical results of surfactant-polymer combination flooding in middle-scale oilfields

Scheme	Time cost	Computational steps	Time cost of the pressure	Balance error (two phases)	Balance error (components)
New algorithm	15213.343	5520	12929.656	10^{-6}	10^{-12}

Table 3: Numerical results of polymer flooding in large-scale oilfields

Scheme	Time cost	Computational steps	Time cost of the pressure	Balance error (two phases)	Balance error (components)
New algorithm	30426.28	6106	20214.90	10^{-6}	10^{-14}

Table 4: Numerical results of surfactant-polymer flooding

Scheme	Time cost	Computational steps	Time cost of the pressure	Balance error (two phases)	Balance error (components)
New algorithm	48877.72	6894	33276.23	10^{-6}	10^{-14}

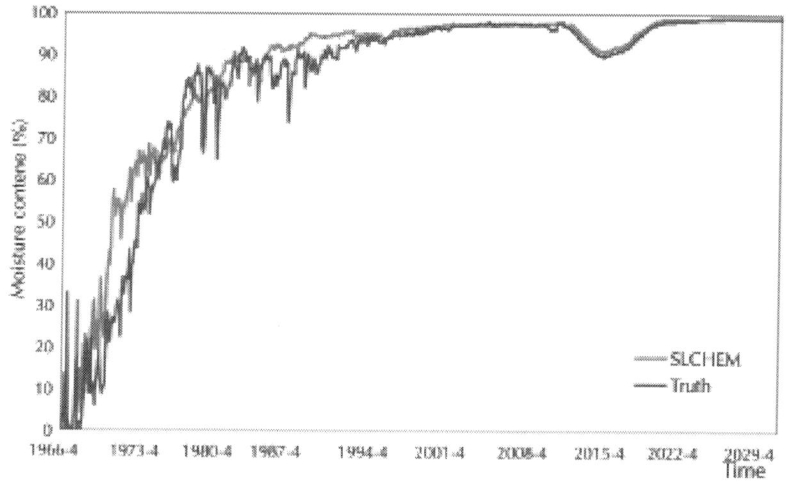

Figure 3: Curve of moisture content simulation of produced oil of two components combination flooding in middle-scale oilfields.

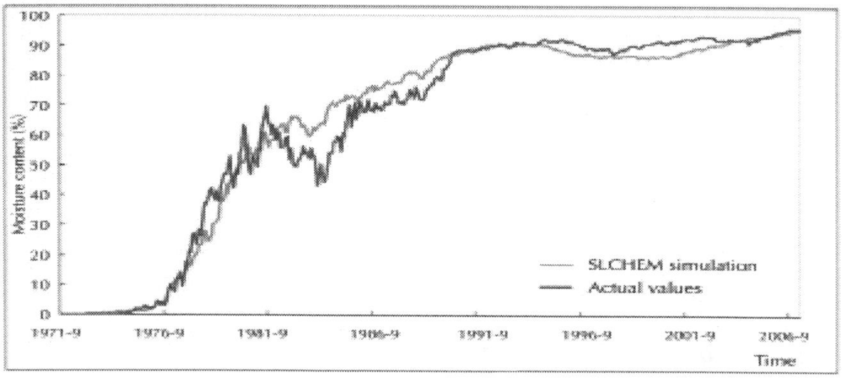

Figure 4: Curve of moisture content simulation of produced oil in large-scale oilfields.

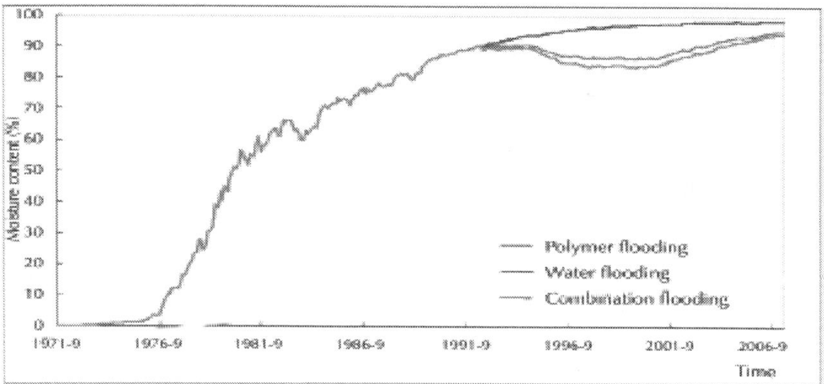

Figure 5: Curve of moisture content simulation of produced oil of surface active agent and combination flooding.

NUMERICAL ANALYSIS OF THE MODEL PROBLEM

In this section theoretical analysis is discussed for the numerical simulation of the polymer of porous media. For simplification and without loss of generality, a simple model is analyzed here, a three-dimensional multi-components compressible displacement [1] [2] [12] - [16] . It is a nonlinear system coupled with partial differential equations with initial values and boundary values described as follows

$$d(c)\frac{\partial p}{\partial t}+\nabla\cdot\boldsymbol{u}=q(x,t),\ x=(x_1,x_2,x_3)^{\mathrm{T}}\in\Omega,\ t\in J=(0,T],\tag{29a}$$

$$\boldsymbol{u}=-a(c)\nabla p,\ x\in\Omega,\ t\in J,\tag{29b}$$

$$\Phi(x)\frac{\partial c_\alpha}{\partial t}+b_\alpha(c)\frac{\partial p}{\partial t}+\boldsymbol{u}\cdot\nabla c_\alpha-\nabla\cdot(D\nabla c_\alpha)=g(x,t,c_\alpha),\ x\in\Omega,\ t\in J,\alpha=1,2,\cdots,n_c-1.\tag{30}$$

Where P(x, t) is the pressure of the mixture, $c_\alpha(x,t)$ is the concentration of the α-th component ($\alpha=1,2,\cdots,n_c-1$)and n_c is the number of components. Because of $\sum_{\alpha=1}^{n_c} c_\alpha(x,t)=1$, the only n_c-1 components are independent. Let $c(x,t)=(c_1(x,t),\cdots c_{nc-1}(x,t))^T$ be a vector function of component saturation, $d(c)=\Phi(x)\sum z_\alpha c_\alpha$, $\Phi(x)$ be the porosity, z_α represent a compressible constant of α-th component, u be the Darcy velocity. $\alpha(c)=k(x)\mu(c)^{-1}$, where k(x) is the permeability, $\mu(c)$ is the fluid viscosity,

And

$$b_\alpha(c)=\Phi c_\alpha\left\{z_\alpha-\sum_{j=1}^{n_c}z_jc_j\right\},$$

D=D(x) is the convection coefficient. The pressure p(x, t) and the vector function of saturations c(x, t) are to be computed later.

Impermeable boundary value conditions:

$$\boldsymbol{u}\cdot\gamma=0,\ x\in\partial\Omega,\quad(D\nabla c_\alpha-c_\alpha\boldsymbol{u})\cdot\gamma=0,\ x\in\partial\Omega,\ t\in J,\alpha=1,2,\cdots,n_c-1,\tag{31}$$

where γ is the outer normal vector of the boundary $\partial\Omega$ of Ω. Initial value conditions:

$$p(x,0)=p_0(x),\ x\in\Omega,\quad c_\alpha(x,0)=c_{\alpha,0}(x),\ x\in\Omega,\ t\in J,\alpha=1,2,\cdots,n_c-1. \tag{32}$$

For convenience, the computational domain is taken as $\Omega=\{[0,1]\}^3$ and the problem is supposed to be Ω-peri odic, and the impermeable boundary value conditions are dropped. Let $h=1/N$, $X_{ijk}=(i_h,j_h,k_h)^T$, $t^n=n\Delta t$ and $W^n_{ijk}=W(X_{ijk},t^n)$, and take

$$A^n_{i+1/2,jk}=\left[a\left(X_{ijk},C^n_{ijk}\right)+a\left(X_{i+1,jk},C^n_{i+1,jk}\right)\right]/2, \tag{33a}$$

$$\delta_{\bar{x}_1}\left(A^n\delta_{x_1}P^{n+1}\right)_{ijk}=h^{-2}\left[A^n_{i+1/2,jk}\left(P^{n+1}_{i+1,jk}-P^{n+1}_{ijk}\right)-A^n_{i-1/2,jk}\left(P^{n+1}_{ijk}-P^{n+1}_{i-1,jk}\right)\right], \tag{33b}$$

Then $A^n_{i,j+1/2,k}$, $A^n_{ij,k+1/2}$ and other notations can be defined similarly.

An implicit fractional finite difference algorithm for the diffusion Equation (29),

$$d\left(C^n_{ijk}\right)\frac{P^{n+1/3}_{ijk}-P^n_{ijk}}{\Delta t}=\delta_{\bar{x}_1}\left(A^n\delta_{x_1}P^{n+1/3}\right)_{ijk}+\delta_{\bar{x}_2}\left(A^n\delta_{x_2}P^n\right)_{ijk}+\delta_{\bar{x}_3}\left(A^n\delta_{x_3}P^n\right)_{ijk}+q\left(X_{ijk},t^{n+1}\right).\ 1\le i\le N. \tag{34a}$$

$$d\left(C^n_{ijk}\right)\frac{P^{n+2/3}_{ijk}-P^{n+1/3}_{ijk}}{\Delta t}=\delta_{\bar{x}_2}\left(A^n\delta_{x_2}\left(P^{n+2/3}-P^n\right)\right)_{ijk},\quad 1\le j\le N, \tag{34b}$$

$$d\left(C^n_{ijk}\right)\frac{P^{n+1}_{ijk}-P^{n+2/3}_{ijk}}{\Delta t}=\delta_{\bar{x}_3}\left(A^n\delta_{x_3}\left(P^{n+1}-P^n\right)\right)_{ijk},\quad 1\le k\le N. \tag{34c}$$

Compute the Darcy velocity $U=(U_1,U_2,U_3)^T$ as follows,

$$U^{n+1}_{1,ijk}=-\frac{1}{2}\left[A^n_{i+1/2,jk}\frac{P^{n+1}_{i+1,jk}-P^{n+1}_{ijk}}{h}+A^n_{i-1/2,jk}\frac{P^{n+1}_{ijk}-P^{n+1}_{i-1,jk}}{h}\right], \tag{35}$$

$U^{n+1}_{2,ijk}$, $U^{n+1}_{3,ijk}$ can be obtained analogously.

An implicit fractional upwind difference method for the saturation Equation (30)

$$\Phi_{ijk}\frac{C_{\alpha,ijk}^{n+1/3}-C_{\alpha,ijk}^{n}}{\Delta t}=\delta_{\bar{x}_1}\left(D\delta_{x_1}C_{\alpha}^{n+1/3}\right)_{ijk}+\delta_{\bar{x}_2}\left(D\delta_{x_2}C_{\alpha}^{n}\right)_{ijk}+\delta_{\bar{x}_3}\left(D\delta_{x_3}C_{\alpha}^{n}\right)_{ijk}-b_{\alpha}\left(C_{\alpha,ijk}^{n}\right)\frac{P_{ijk}^{n+1}-P_{ijk}^{n}}{\Delta t}$$
$$+g\left(X_{ijk},t^{n+1},C_{\alpha,ijk}^{n}\right) \qquad ,1\le i\le N,\alpha-1,2,\cdots,n_c-1, \tag{36a}$$

$$\Phi_{ijk}\frac{C_{\alpha,ijk}^{n+2/3}-C_{\alpha,ijk}^{n+1/3}}{\Delta t}=\delta_{\bar{x}_2}\left(D\delta_{x_2}\left(C_{\alpha,ijk}^{n+2/3}-C_{\alpha}^{n}\right)\right)_{ijk}, \qquad 1\le j\le N,\alpha=1,2,\cdots,n_c-1, \tag{36b}$$

$$\Phi_{ijk}\frac{C_{\alpha,ijk}^{n+1}-C_{\alpha,ijk}^{n+2/3}}{\Delta t}=\delta_{\bar{x}_3}\left(D\delta_{x_3}\left(C_{\alpha,ijk}^{n+1}-C_{\alpha}^{n}\right)\right)_{ijk}-\sum_{\beta=1}^{3}\delta_{U_{\beta}^{n},x_{\beta}}C_{\alpha,ijk}^{n+1}, \qquad 1\le k\le N,\alpha=1,2,\cdots,n_c-1, \tag{36c}$$

Where

$$\delta_{U_{\beta}^{n},x_{\beta}}C_{\alpha,ijk}^{n+1}=U_{\beta,ijk}^{n}\left\{H\left(U_{\beta,ijk}^{n}\right)\delta_{\bar{x}_{\beta}}+\left(1-H\left(U_{\beta,ijk}^{n}\right)\right)\delta_{x_{\beta}}\right\}C_{\alpha,ijk}^{n+1}, \ \alpha=1,2,\cdots,n_c-1. \ \ H(z)=\begin{cases}1, & z\ge 0\\ 0, & z<0\end{cases}$$

Initial value conditions:

$$P_{ijk}^{0}=p_0\left(X_{ijk}\right), \ \ C_{\alpha,ijk}^{0}=c_{\alpha 0,ijk}, \ \ 1\le i,j,k\le N, \ \ \alpha=1,2,\cdots,n_c-1. \tag{37}$$

The program of implicit upwind fractional difference method runs as follows. Given $\{P_{ijk}^{n},C_{\alpha,ijk}^{n},\alpha=1,2,\cdots,n_c-1\}$, the values $\{P_{ijk}^{n+1/3}\}$ of transition layer is computed by using speedup method in x_1-direction by (34a), $\{P_{ijk}^{n+2/3}\}$ is obtained by (34b), and the solution of pressure $\{P_{ijk}^{n+1}\}$ is solved by (34c). The values of Darcy velocity $\{U_{ijk}^{n+1}\}$ are computed by (35). Secondly, the values $\{C_{\alpha,ijk}^{n+1/3}\}$ of transition layer is computed by using speedup method in x_1-direction by (36a), $\{C_{\alpha,ijk}^{n+2/3}\}$ is obtained by (36b), and the solution of concentration $\{C_{\alpha,ijk}^{n+1}\}$ is solved by (36c). Note that the problem is positive definite, the solution of (34)-(36) exists and is sole.

For the model (29)-(32), applying variation technique, energy analysis, decomposition of high order difference operators and theory and technique of product communitivity we can get error estimates in L^2 of the implicit fractional upwind difference method stated in the following theorem. The computational algorithms discussed in this paper is based on the mathematical and mechanical considerations.

Theorem Suppose that the exact solutions of (29)-(32) are sufficiently smooth, and discrete solutions are computed by an implicit fractional step upwind difference algorithm (34)-(36). It holds

$$\|p-P\|_{\bar{L}^\infty(J;h^1)}+\sum_{\alpha=1}^{n_c-1}\|c_\alpha-C_\alpha\|_{\bar{L}^\infty(J;h^1)}+\|d_t(p-P)\|_{\bar{L}^2(J;l^2)}+\sum_{\alpha=1}^{n_c-1}\|d_t(c_\alpha-C_\alpha)\|_{\bar{L}^2(J;l^2)}\le M^*\{\Delta t+h\}. \quad (38)$$

where the constant M* is dependent on p(x,t), $c_\alpha(x,t)(\alpha=1,2,\cdots,n_c-1)$ and their derivatives.

CONCLUSIONS AND DISCUSSION

Theory, method and application of numerical simulation of high temperature, high salt, complicated geology and large scale oilfields and complicated chemical flooding of flow mechanics in porous media are discussed in this paper consisting of several sections. Summary is stated about our project in Section 1. Mathematical model of permeation fluid mechanics is presented in view of petroleum geology, geochemistry, and computational permeation fluid mechanics Section 2. An implicit refined fractional step combined with upwind difference numerical algorithm based on upstream sequence, is structured in Section 3. A type of software applicable in major industries has been accomplished, mostly carried out the spacial step of ten meters, ten thousands of nodes and tens of years simulation period in Section 4. Some experimental tests taking place successfully in major oil fields such as Shengli Oilfield, are illustrated in Section 5. Numerical analysis proceeds for the model problem and precise theoretical results are stated on mathematical and mechanical consideration in Section 6.

ACKNOWLEDGMENTS

The authors express their deep appreciation to prof. J. Douglas Jr, prof. R. E. Ewing and prof. Jiang Lishang for their many helpful suggestions in the serial of research of chemical production. This work is supported by National climbing plan (Grant No. 1990328), National Tackling Key Project (Grant No. 20050200069), Natural Science Foundation of China (Grant Nos. 11101244, 11271231, 10771124, 10372052), National Doctoral Foundation (Grant No. 20030422047) and Natural Science Foundation of Shandong Province (Grant Nos. ZR2011AM015, ZR2009AQ012), National Major Programs for Science and Technology Development (2011ZX05011-004, 2011ZX05052).

REFERENCES

1. Ewing, R.E, Yuan, Y.R. and Li, G. (1989) Finite Element for Chemical-Flooding Simulation.*Proceeding of the 7th International Conference Finite Element Method in Flow Problems,* The University of Alabama in Huntsville, Huntsville, 1264-1271.
2. Yuan, Y.R. (2013) Theory and Application of Numerical Simulation Energy Sources. *Basis of Numerical Simulation of Chemical Production* (*Tertiary Oil Recovery*), Science Press, Beijing, 257-304.
3. Institute of Mathematics, Shandong University, Exploration Institute of Daqing Petroleum Administration (1995) Research and Application of the Polymer Flooding Software. Summary of "Eighth-Five" National Key Science and Tech- nology Program, Grant No. 85-203-01-08.
4. Yuan, Y.R. (1993) The Characteristic Finite Difference Method for Enhanced Oil Recovery Simulation and L^2 Estimates. *Science in China* (*A*), 11, 1296-1307.
5. Yuan, Y.R. (1994) The Characteristic Mixed Finite Element Method and Analysis for Two Dimensional Chemical Flooding

Reservoir Simulation. *Acta Mathematicae Applicatae Sinica*, 1, 118-131.

6. Yuan, Y.R. (1993) The Characteristic Mixed Finite Element Method for Enhanced Oil Recovery Simulation and Op- timal Order L^2 Error Estimate. *Chinese Science Bulletin*, 21, 1761-1766.
7. China National Petroleum Corporation (1995) Evaluation Report of Executive Condition of "Eighth-Five" National Key Science and Technology Program. Grant No. 85-203-01-08.
8. Yuan, Y.R., Yang, D.P., Qi, L.Q., *et al.* (1998) Research on Algorithms of Applied Software of the Polymer. In: Gang, Q.L., Ed., *Proceeding on Chemical Flooding*, Petroleum Industry Press, Beijing, 246-253.
9. Yuan, Y.R. (2000) Finite Element Method and Analysis for Chemical Flow Simulation.*System Science and Mathematical Sciences*, 3, 302-308.
10. Institute of Mathematics, Shandong University, Exploration Institute of Daqing Petroleum Administration (2006) Modification of Solving Mathematical Models of the Polymer and Improvement of Reservoir Description.
11. Institute of Mathematics, Shandong University, Shengli Oilfield Branch, China Petroleum and Chemical Corporation (2011) Research on Key Technology of High Temperature and High Salinity Chemical Agent Displacement. 83-106.
12. Ewing, R.E. (1983) The Mathematics of Reservoir Simulation. SIAM, Philadelphia.http://dx.doi.org/10.1137/1.9781611971071
13. Yuan, Y.R. (2002) The Upwind Difference Method for Compressible Two-Phase Displacement Problem. *Acta Mathematicae Applicatae Sinica*, 3, 484-496.
14. Yuan, Y.R. (1999) The Characteristic Finite Difference Fractional Steps Method for Compressible Two-Phase Displacement Problem. *Science in China* (*A*), 1, 48-57.http://dx.doi.org/10.1007/BF02872049
15. Yuan, Y.R. (2003) The Upwind Finite Difference Fractional

Steps Method for Two-Phase Compressible Flow in Porous Media. *Numerical Methods for Partial Differential Equations*, 19, 67-88. http://dx.doi.org/10.1002/num.10036

16. Yuan, Y.R. (2001) Characteristic Finite Difference Fractional Steps Method for Three-Dimensional Compressible Multi-Component Displacement Problem. *Acta Mathematicae Applicatae Sinica*, 2, 242-249.

Citations

CHAPTER 1

Don B. Fox, Daniel Sutter, Koenraad F. Beckers, Maciej Z. Lukawski, Donald L. Koch, Brian J. Anderson, Jefferson W. Tester, Sustainable heat farming: Modeling extraction and recovery in discretely fractured geothermal reservoirs, Geothermics, Volume 46, April 2013, Pages 42-54, ISSN 0375-6505, http://dx.doi.org/10.1016/j.geothermics.2012.09.001.

CHAPTER 2

H. Mohammadi, A. Khaksar Manshad, G.H. Montazeri, A reservoir simulation approach for modeling of naturally fractured reservoirs, Egyptian Journal of Petroleum, Volume 21, Issue 2, December 2012, Pages 125-133, ISSN 1110-0621, http://dx.doi.org/10.1016/j.ejpe.2012.11.002.

CHAPTER 3

Jihoon Kim, Eric Sonnenthal, Jonny Rutqvist, A sequential implicit algorithm of chemo-thermo-poro-mechanics for fractured geothermal reservoirs, Computers & Geosciences, Volume 76, March 2015, Pages 59-71, ISSN 0098-3004, http://dx.doi.org/10.1016/j.cageo.2014.11.009.

CHAPTER 4

V. Vishal, P.G. Ranjith, S.P. Pradhan, T.N. Singh, Permeability of sub-critical carbon dioxide in naturally fractured Indian bituminous coal at a range of down-hole stress conditions, Engineering Geology, Volume 167, 17 December 2013, Pages 148-156, ISSN 0013-7952, http://dx.doi.org/10.1016/j.enggeo.2013.10.007.

CHAPTER 5

Albert Genter, Keith Evans, Nicolas Cuenot, Daniel Fritsch, Bernard Sanjuan, Contribution of the exploration of deep crystalline fractured reservoir of Soultz to the knowledge of enhanced geothermal systems (EGS), Comptes Rendus Geoscience, Volume 342, Issues 7–8, July–August 2010, Pages 502-516, ISSN 1631-0713, http://dx.doi.org/10.1016/j.crte.2010.01.006.

CHAPTER 6

Joshua Taron, Derek Elsworth, Thermal–hydrologic–mechanical–

chemical processes in the evolution of engineered geothermal reservoirs, International Journal of Rock Mechanics and Mining Sciences, Volume 46, Issue 5, July 2009, Pages 855-864, ISSN 1365-1609, http://dx.doi.org/10.1016/j.ijrmms.2009.01.007.

CHAPTER 7

Yuan, Y. , Cheng, A. , Yang, D. and Li, C. (2014) Theory and Application of Numerical Simulation of Chemical Flooding in High Temperature and High Salt Reservoirs. International Journal of Geosciences, 5, 956-970. doi: 10.4236/ijg.2014.59082.

Index

I

L

M

N

O

P

R

T

V